Author:
Dana M. Barry, Ph.D.
Editor for the Center for
Advanced Materials
Processing (CAMP)
Clarkson University

Illustrator:
Kathy Bruce

Editors:
Evan D. Forbes, M.S. Ed.
Walter Kelly, M.A.

Senior Editor:
Sharon Coan, M.S. Ed.

Art Director:
Darlene Spivak

Product Manager:
Phil Garcia

Imaging:
Rick Chacon

Research:
Bobbie Johnson

Publishers:
Rachelle Cracchiolo, M.S. Ed.
Mary Dupuy Smith, M.S. Ed.

Hands-On Minds-On Science
Easy Chemistry

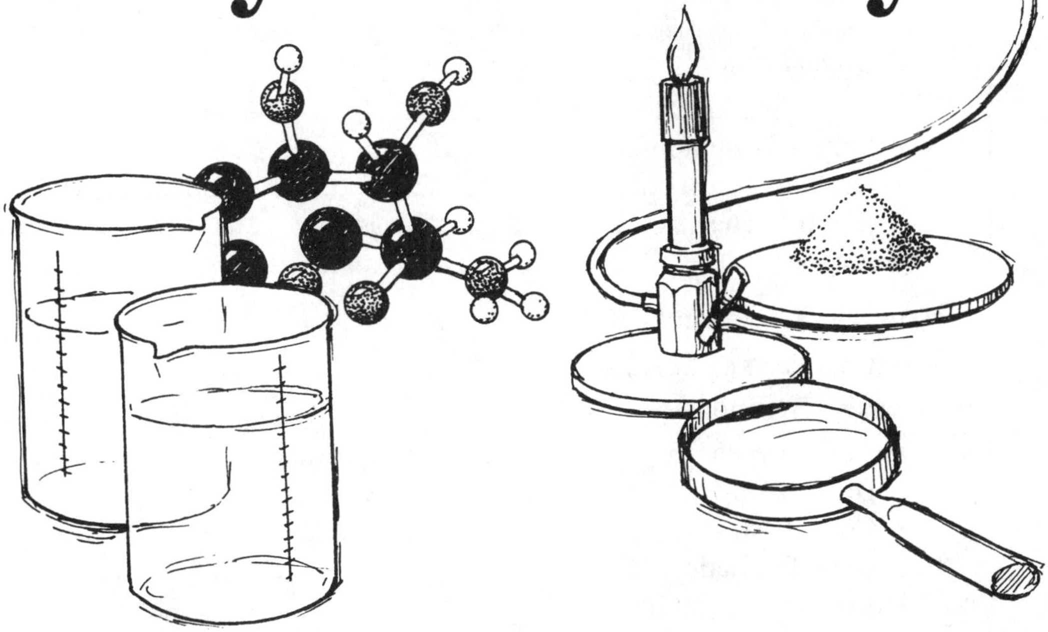

Teacher Created Materials, Inc.
P.O. Box 1040
Huntington Beach, CA 92647
©*1994 Teacher Created Materials, Inc.*
Made in U.S.A.

ISBN-1-55734-648-8

The classroom teacher may reproduce copies of materials in this book for classroom use only. The reproduction of any part for an entire school or school system is strictly prohibited. No part of this publication may be transmitted, stored, or recorded in any form without written permission from the publisher.

Table of Contents

Introduction 4
The Scientific Method 5
Science-Process Skills 7
Organizing Your Unit 9

What are the Parts of a Chemical?
Just the Facts 11
Hands-On Activities
- The Smallest Part 12
- Name 15
- Holding Together 17
- Hidden Water 20
- Group Leader 22
- Disappearing Act 24

What Are the Properties of a Chemical?
Just the Facts 26
Hands-On Activities
- Appearance 27
- More than Meets the Eye 29
- Solution Dilution 32
- Density 35
- Pile It Higher 37
- Let's Race 39

How Do Chemicals React?
Just the Facts 41
Hands-On Activities
- Add Oxygen 42
- Blow It Up 44
- Color Change 46
- Solid Formation 48
- Delicious Combination 51
- Changing Partners 53

Where Are Chemicals Found?
Just the Facts 56
Hands-On Activities
- Human Body 57
- Water 59
- Air 61
- Rocks and Minerals 63
- Food and Beverage 66
- Everywhere 69

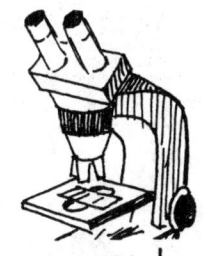

Table of Contents *(cont.)*

Curriculum Connections

- Language Arts 71
- Social Studies 72
- Physical Education 73
- Math 74
- Art...75
- Music 76

Station-to-Station Activities

- Observe..77
- Communicate 78
- Compare 79
- Order..80
- Categorize..81
- Relate 82
- Infer 83
- Apply 84

Management Tools

- Science Safety 85
- Chemistry Journal 86
- Chemistry Observation Area..90
- Assessment Form 91
- Science Award 92

Glossary 93

Bibliography 95

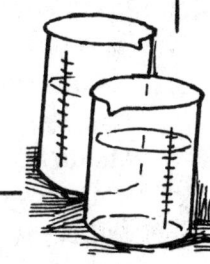

Introduction

What Is Science?

What is science to young children? Is it something that they know is a part of their world? Is it a textbook in the classroom? Is it a tadpole changing into a frog? A sprouting seed, a rainy day, a boiling pot, a turning wheel, a pretty rock, or a moonlit sky? Is science fun and filled with wonder and meaning? What is science to a child?

Science offers you and your eager students opportunities to explore the world around you, and make connections between the things you experience. The world becomes your classroom, and you, the teacher, a guide.

Science can, and should, fill children with wonder. It should cause them to be filled with questions and the desire to discover the answers to their questions. And, once they have discovered answers, they should be actively seeking new questions to answer.

The books in this series give you and the students in your classroom the opportunity to learn from the whole of your experience—the sights, sounds, smells, tastes, and touches, as well as what you read, write about, and do. This whole-science approach allows you to experience and understand your world as you explore science concepts and skills together.

What is Chemistry?

Chemistry is the study of substances, natural and artificial. Chemists are the people who study these substances under a variety of conditions to see how they act or react. We find chemicals everywhere—in the food we eat, the clothes we wear, and the air we breathe. Because chemicals and chemical substances are so common, it is important to know what they are, what they do, and how they can be used. Approximately 1.5 million years ago the first known chemical reaction, fire, was discovered. Through the use of fire, people were able to change the properties of substances. It was used to cook food, to harden pottery, and to smelt metal ore. Since the discovery of the first known chemical reaction, many more are used today, and others are still waiting to be discovered. The study of chemistry has been divided into several branches to aid in our discovery and use of substances. The branches of chemistry are these: analytical, applied, bio, inorganic, physical, polymer, and synthetic. Welcome to the wonderful world of chemistry.

The Scientific Method

The scientific method is a creative and systematic process for proving or disproving a given question, following an observation. When scientists use the scientific method, a basic set of guiding principles and procedures is followed in order to obtain new knowledge about our universe. This method will be described in the paragraphs that follow.

It is easy to teach the scientific method! Just follow these simple steps:

1 Make an OBSERVATION.

The teacher presents a situation, gives a demonstration, or reads background material that interests students and prompts them to ask questions. Or students can make observations and generate questions on their own as they study a topic.

Example: A blown-up balloon.

2 Select a QUESTION to investigate.

In order for students to select a question for a scientific investigation, they will have to consider the materials they have or can get, as well as the resources (books, magazines, people, etc.) actually available to them. You can help them make an inventory of their materials and resources, either individually or as a group.

Tell students that in order to successfully investigate the questions they have selected, they must be very clear about what they are asking. Discuss effective questions with your students. Depending upon their level, simplify the question or make it more specific.

Example: How can one detect a chemical change?

3 Make a PREDICTION (Hypothesis).

Explain to students that a hypothesis is a good guess about what the answer to a question will probably be. But they do not want to make just any arbitrary guess. Encourage students to predict what they think will happen and why.

In order to formulate a hypothesis, students may have to gather more information through research.

Have students practice making hypotheses with questions you give them. Tell them to pretend they have already done their research. You want them to write each hypothesis so it follows these rules:

1. It is to the point.
2. It tells what will happen, based on what the question asks.
3. It follows the subject/verb relationship of the question.

Example: I think when the chemicals react, the balloon will pop.

The Scientific Method *(cont.)*

4. Develop a **PROCEDURE** to test the hypothesis.

The first thing students must do in developing a procedure (the test plan) is to determine the materials they will need.

They must state exactly what needs to be done in step-by-step order. If they do not place their directions in the right order, or if they leave out a step, it becomes difficult for someone else to follow their directions. A scientist never knows when other scientists will want to try the same experiment to see if they end up with the same results!

Example: Mixing baking soda and vinegar together, will cause the balloon to inflate.

5. Record the **RESULTS** of the investigation in written and picture form.

The results (data collected) of a scientific investigation are usually expressed two ways—in written form and in picture form. Both are summary statements. The written form reports the results with words. The picture form (often a chart or graph) reports the results so the information can be understood at a glance.

Example: The results of the investigation can be recorded on a data-capture sheet provided (page 45).

6. State a **CONCLUSION** that tells what the results of the investigation mean.

The conclusion is a statement which tells the outcome of the investigation. It is drawn after the student has studied the results of the experiment, and it interprets the results in relation to the stated hypothesis. A conclusion statement may read something like either of the following: "The results show that the hypothesis is supported," or "The results show that the hypothesis is not supported." Then restate the hypothesis if it was supported or revise it if it was not supported.

Example: The hypothesis that stated "when the chemicals react, the balloon will pop" is supported (or not supported).

7. Record **QUESTIONS, OBSERVATIONS**, and **SUGGESTIONS** for future investigations.

Students should be encouraged to reflect on the investigations that they complete. These reflections, like those of professional scientists, may produce questions that will lead to further investigations.

Example: What was the gas created when the baking soda and vinegar were combined?

Science-Process Skills

Even the youngest students blossom in their ability to make sense out of their world and succeed in scientific investigations when they learn and use the science-process skills. These are the tools that help children think and act like professional scientists.

The first five process skills on the list below are the ones that should be emphasized with young children, but all of the skills will be utilized by anyone who is involved in scientific study.

Observing

It is through the process of observation that all information is acquired. That makes this skill the most fundamental of all the process skills. Children have been making observations all their lives, but they need to be made aware of how they can use their senses and prior knowledge to gain as much information as possible from each experience. Teachers can develop this skill in children by asking questions and making statements that encourage precise observations.

Communicating

Humans have developed the ability to use language and symbols which allow them to communicate not only in the "here and now" but also over time and space as well. The accumulation of knowledge in science, as in other fields, is due to this process skill. Even young children should be able to understand the importance of researching others' communications about science and the importance of communicating their own findings in ways that are understandable and useful to others. The chemistry journal and the data-capture sheets used in this book are two ways to develop this skill.

Comparing

Once observation skills are heightened, students should begin to notice the relationships between things that they are observing. Comparing means noticing similarities and differences. By asking how things are alike and different or which is smaller or larger, teachers will encourage children to develop their comparison skills.

Ordering

Other relationships that students should be encouraged to observe are the linear patterns of seriation (order along a continuum: e.g., rough to smooth, large to small, bright to dim, few to many) and sequence (order along a time line or cycle). By ranking graphs, time lines, cyclical and sequence drawings, and by putting many objects in order by a variety of properties, students will grow in their abilities to make precise observations about the order of nature.

Categorizing

When students group or classify objects or events according to logical rationale, they are using the process skill of categorizing. Students begin to use this skill when they group by a single property such as color. As they develop this skill, they will be attending to multiple properties in order to make categorizations; the animal classification system, for example, is one system students can categorize.

Science-Process Skills *(cont.)*

Relating

Relating, which is one of the higher-level process skills, requires student scientists to notice how objects and phenomena interact with one another and the change caused by these interactions. An obvious example of this is the study of chemical reactions.

Inferring

Not all phenomena are directly observable, because they are out of humankind's reach in terms of time, scale, and space. Some scientific knowledge must be logically inferred based on the data that is available. Much of the work of paleontologists, astronomers, and those studying the structure of matter is done by inference.

Applying

Even very young, budding scientists should begin to understand that people have used scientific knowledge in practical ways to change and improve the way we live. It is at this application level that science becomes meaningful for many students.

Organizing Your Unit

Designing a Science Lesson

In addition to the lessons presented in this unit, you will want to add lessons of your own, lessons that reflect the unique environment in which you live, as well as the interests of your students. When designing new lessons or revising old ones, try to include the following elements in your planning:

Question

Pose a question to your students that will guide them in the direction of the experiment you wish to perform. Encourage all answers, but you want to lead the students towards the experiment you are going to be doing. Remember, there must be an observation before there can be a question. (Refer to The Scientific Method, pages 5-6.)

Setting the Stage

Prepare your students for the lesson. Brainstorm to find out what students already know. Have children review books to discover what is already known about the subject. Invite them to share what they have learned.

Materials Needed for Each Group or Individual

List the materials each group or individual will need for the investigation. Include a data-capture sheet when appropriate.

Procedure

Make sure students know the steps to take to complete the activity. Whenever possible, ask them to determine the procedure. Make use of assigned roles in group work. Create (or have your students create) a data-capture sheet. Ask yourself, "How will my students record and report what they have discovered? Will they tally, measure, draw, or make a checklist? Will they make a graph? Will they need to preserve specimens?" Let students record results orally, using a video or audio tape recorder. For written recording, encourage students to use a variety of paper supplies such as poster board or index cards. It is also important for students to keep a journal of their investigation activities. Journals can be made of lined and unlined paper. Students can design their own covers. The pages can be stapled or be put together with brads or spiral binding.

Extensions

Continue the success of the lesson. Consider which related skills or information you can tie into the lesson, like math, language arts skills, or something being learned in social studies. Make curriculum connections frequently and involve the students in making these connections. Extend the activity, whenever possible, to home investigations.

Closure

Encourage students to think about what they have learned and how the information connects to their own lives. Prepare journals using "Chemistry Journal" directions on page 86. Provide an ample supply of blank and lined pages for students to use as they complete the "Closure" activities. Allow time for students to record their thoughts and pictures in their journals.

The Big Why

The explanation behind the experience is provided.

Organizing Your Unit (cont.)

Structuring Student Groups for Scientific Investigations

Using cooperative learning strategies in conjunction with hands-on and discovery learning methods will benefit all the students taking part in the investigation.

Cooperative Learning Strategies

1. In cooperative learning, all group members need to work together to accomplish the task.
2. Cooperative learning groups should be heterogenous.
3. Cooperative learning activities need to be designed so that each student contributes to the group and individual group members can be assessed on their performance.
4. Cooperative learning teams need to know the social as well as the academic objectives of a lesson.

Cooperative Learning Groups

Groups can be determined many ways for the scientific investigations in your class. Here is one way of forming groups that has proven to be successful in intermediate classrooms.

- **The Team Leader**—scientist in charge of reading directions and setting up equipment.
- **The Chemist**—scientist in charge of carrying out directions (can be more than one student).
- **The Stenographer**—scientist in charge of recording all of the information.
- **The Transcriber**—scientist who translates notes and communicates findings.

If the groups remain the same for more than one investigation, require each group to vary the people chosen for each job. All group members should get a chance to try each job at least once.

Using Centers for Scientific Investigations

Set up stations for each investigation. To accommodate several groups at a time, stations may be duplicated for the same investigation. Each station should contain directions for the activity, all necessary materials (or a list of materials for investigators to gather), a list of words (a word bank) which students may need for writing and speaking about the experience, and any data-capture sheets or needed materials for recording and reporting data and findings.

Station-to-Station Activities are on pages 77-84. Model and demonstrate each of the activities for the whole group. Have directions at each station. During the modeling session, have a student read the directions aloud while the teacher carries out the activity. When all students understand what they must do, let small groups conduct the investigations at the centers. You may wish to have a few groups working at the centers while others are occupied with other activities. In this case, you will want to set up a rotation schedule so all groups have a chance to work at the centers.

Assign each team to a station, and after they complete the task described, help them rotate in a clockwise order to the other stations. If some groups finish earlier than others, be prepared with another unit-related activity to keep students focused on main concepts. After all rotations have been made by all groups, come together as a class to discuss what was learned.

What Are the Parts of a Chemical?

Just the Facts

Chemicals are everywhere. Ask the students if they know what a chemical is. Have them write their chemical descriptions on a piece of paper. Discuss these descriptions in class.

A chemical is a substance that has mass and takes up space. It contains one or more elements and is represented by a formula with the symbols of the elements in it. A chemical's name includes the elements making up the chemical.

Chemicals contain the following parts:

- An **element**...is a substance that cannot be broken down into simpler substances. It is represented by a symbol consisting of letters from the alphabet.

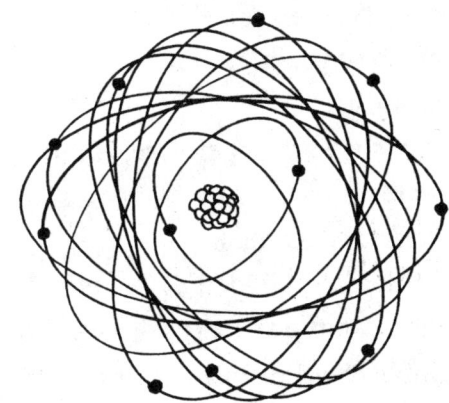

- An **atom**...is the smallest particle that describes an element. The center or nucleus of the atom contains protons (positively charged particles) and neutrons (neutral particles). Electrons (negatively charged particles) move around the center of the atom.

- A **molecule**...is a group of bonded atoms that exists as a separate entity. Chemicals called hydrates are crystals that contain water molecules.

- A **functional group**...is a specific group of atoms that gives characteristic properties to the chemical that they are part of.

- A **bond**...holds a chemical together. Covalent bonds are formed by the sharing of electrons, and ionic bonds are formed by the attraction of oppositely charged ions (charged atoms).

What Are the Parts of a Chemical?

The Smallest Part

Question
What is the smallest part of a chemical?

Setting the Stage

- Discuss the various parts of a chemical with your students. Include *elements, atoms*, and *molecules*. Tell them an element is a substance that cannot be broken down into simpler substances, an atom is the smallest particle that describes an element, and a molecule is a group of bonded atoms that exists as a separate entity.

- Describe with students the *Bohr model* of the atom. Tell them that the center or the nucleus of an atom contains positively charged particles called *protons* and neutral particles called *neutrons*. In the Bohr model, the *electrons* (negatively charged particles of the atom) revolve around the nucleus in concentric orbits.

Materials Needed for Each Group

- piece of paper
- scissors
- thin strip of copper wire, about 6" (15 cm) long
- water
- beaker (100 mL)
- graduated cylinders (100 mL, 50 mL, and 10 mL)
- data-capture sheet (page 14), one per student

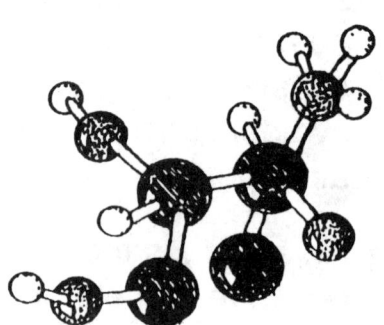

Procedure *(Student Instructions)*

1. Fold a piece of paper in half and carefully cut it on the fold.
2. Take one piece of the cut paper and again fold it in half and cut it on the fold.
3. Continue to fold and cut the paper in half until it can no longer be folded and cut.
4. Keep track of the number of times the paper has been folded and cut.
5. Determine the size of the smallest piece of paper compared to the whole. Each time the paper is folded and cut, multiply by one-half. If the paper is folded and cut four times, the smallest piece will be one-sixteenth of the whole.
6. Repeat steps 1-5 to prepare a small piece of copper wire and determine its size compared to the whole.
7. Using a graduated cylinder, measure 100 mL of water and pour it into a beaker.
8. Using a 50 mL graduated cylinder, remove half the water from the beaker. The beaker should contain 50mL water.
9. Again remove half the water from the beaker. The beaker should contain 25 mL water.
10. Continue the process of removing half the water from the beaker for as many times as you are able to measure the volume in mL.
11. Keep track of the number of times that half the water is removed from the beaker.
12. Determine the smallest amount of water compared to the original 100 mL. Each time the beaker is emptied (half-way), multiply by one-half as was done for the paper and copper wire. Note the number of mL in the smallest amount of measured water.
13. Complete your data-capture sheet.

What Are the Parts of a Chemical?

The Smallest Part *(cont.)*

Extensions

- Have students use magnifying glasses and microscopes to look at very small items (eg., strands of hair).
- Have students form a human Bohr model of one atom of oxygen which has an atomic number of eight and contains eight protons and eight electrons. Have eight students (representing protons) each hold a plus sign and stand together in the center of the room. Have two students (representing electrons) each hold a negative sign and a piece of string to form a concentric circle around the protons. Have six more students (representing electrons) form a second concentric circle around the protons.

Closure

In their chemistry journals, have students write whether they were able to obtain the smallest possible amount of each substance tested. Have them explain their answers.

The Big Why

This activity enhances students' perception of the size of a substance compared to its smallest part. Tell students that they cannot see the smallest part of a substance with the naked eye. One can see atoms with a scanning tunneling microscope (STM).

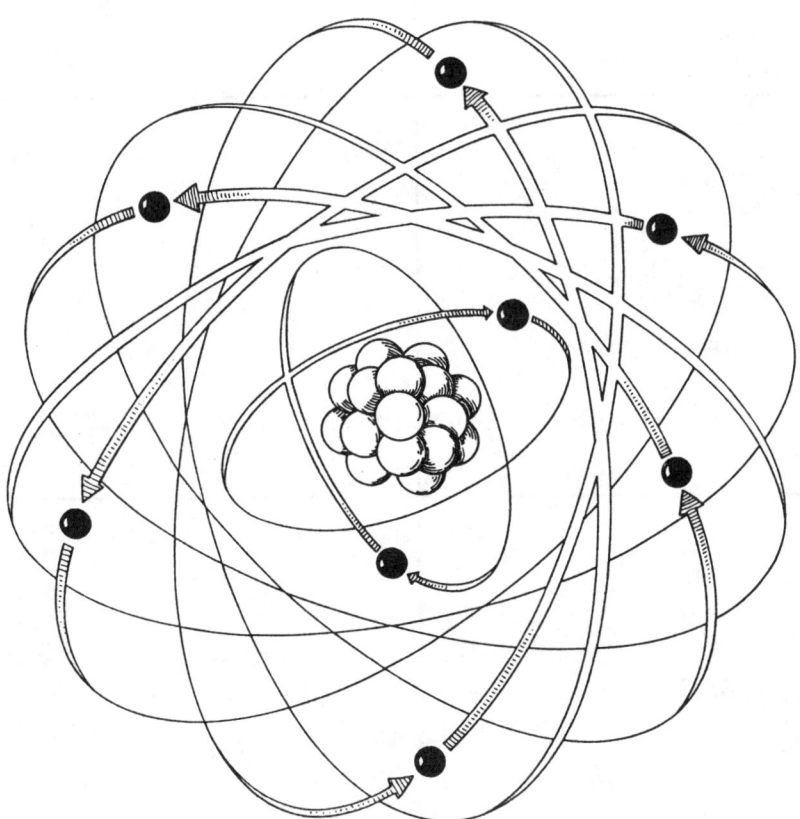

What Are the Parts of a Chemical?

The Smallest Part (cont.)

Complete the chart below. Do your calculations in the space provided.

Item or Chemical	Size of the Smallest Part Compared to the Whole
paper	
copper wire	
water	

#648 Easy Chemistry

What Are the Parts of a Chemical?

Name

Question

What information is found in a chemical's name and formula?

Setting the Stage
- Explain how *symbols* are used to represent elements and how they are used in a chemical formula. Show students the Periodic Table of the Elements.
- Discuss with students the chemical named magnesium sulfate with the formula $MgSO_4 \cdot 7H_2O$. Tell students that H_2O stands for water and that the coefficient seven stands for the number of molecules of water attached to the salt, magnesium sulfate. Tell them that the symbol Mg stands for magnesium, S stands for sulfur, O stands for oxygen, and the subscript stands for the number of atoms of a particular element.
- Give students the symbols for the elements that they will be using in this activity.

Materials Needed for Each Group
- six labeled bottles of various chemicals
- data-capture sheet (page 16), one per student

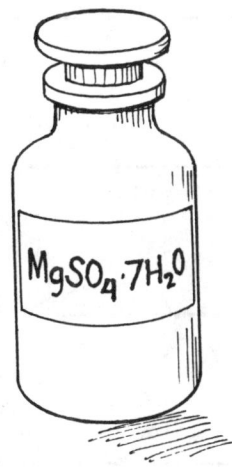

Procedure *(Student Instructions)*
1. Look at a labeled chemical bottle. Note its name and formula and record.
2. Repeat step 1 with the other 5 chemicals.
3. Complete your data-capture sheet.

Extensions
- Have students further analyze the data on their data-capture sheets. Have them group the chemicals according to the common elements—eg., All chemicals containing oxygen would be put in one group. Have students determine the total number of atoms in each chemical.
- Have students look at and analyze other labeled chemicals.

Closure

In their chemistry journals, have students describe the different types of information that they got from the chemical name and formula.

The Big Why

This activity acquaints students with the names of chemicals and substances. Point out to the students that they themselves are identified by names and so are chemicals. A chemical's name tells what elements it contains. Each element is represented by a symbol. A chemical's formula contains the symbols of the elements in it.

©1994 Teacher Created Materials, Inc. #648 *Easy Chemistry*

What Are the Parts of a Chemical?

Name *(cont.)*

After examining the chemical bottles, fill in the data.

Chemical Name	Chemical Formula	Elements Present	Number/Atoms of Element	Molecules of Water

What Are the Parts of a Chemical?

Holding Together

Question
What holds chemicals together?

Setting the Stage
- Discuss *ionic* and *covalent bonding* with students. Point out that covalent bonding is a sharing of electrons and ionic bonding is an attraction of oppositely charged ions.
- Discuss *positive* and *negative ions* with students. Mention that a positive ion is formed when an atom loses an electron and that a negative ion is formed when an atom gains an electron.

Materials Needed for Each Individual
- two magnetic marbles of a different color or two magnets
- straws (six cut in half)
- colored marshmallows
- data-capture sheet (page 19)

Procedure *(Student Instructions)*
1. Hold the two magnetic marbles close together. Note what happens. (Tell students that the magnetic marbles represent oppositely charged ions such as the positive sodium ion and the negative chloride ion in the table salt, sodium chloride.)
2. Pull the marbles apart. Note that energy is required to break the bond.
3. Make a model of hydrogen (H_2), a missile fuel. Let a white marshmallow represent each hydrogen. Let the straw represent the bond between the two hydrogens.
4. Make a model of methane (CH_4), natural gas. Carbon is represented by C and is singly bonded to each of the four hydrogens. Let the straws represent the bonds. Let a white marshmallow represent each hydrogen. Let a colored marshmallow represent carbon.
5. Make a model of ethane (C_2H_4), a gas used in refrigerants and welding. Here a carbon is double bonded to a carbon so that two pairs of electrons are being shared. Each carbon has a single bond to two hydrogens. Again use straws to represent bonds, white marshmallows to represent hydrogen, and colored marshmallows to represent carbon.
6. Complete your data-capture sheet.

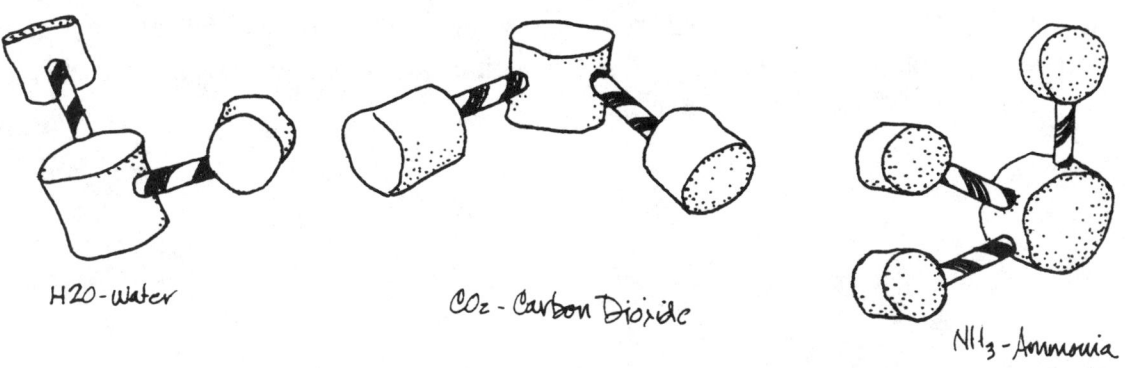

H2O - Water   CO2 - Carbon Dioxide   NH3 - Ammonia

©1994 Teacher Created Materials, Inc.　　　17　　　#648 Easy Chemistry

What Are the Parts of a Chemical?

Holding Together (cont.)

Extensions
- Have students compare the strengths of different materials. Have them rip a piece of paper and piece of cardboard, noting which bond took more energy to break.
- Have students show bond strengths by stretching rubber bands over pairs of nails at different distances apart on a wooden board. Students should measure and record in centimeters (cm) the longest distance each rubber band can be stretched before breaking.

Closure

In your chemistry journal, have students describe the similarities and differences between ionic and covalent bonds.

The Big Why
- **Ionic Bonds:** Each electron has a charge of negative one, and each proton has a charge of positive one. Normally atoms are neutral, having an equal number of protons and electrons. Therefore, the negative and positive charges cancel each other out. The sodium atom (Na) has 11 electrons and 11 protons (11 negatives and 11 positives) resulting in an overall charge of 0. Sodium, a metal, tends to lose an electron to form a positive ion (or charged atom) with ten electrons and 11 protons. The extra proton gives the sodium a charge of positive one. Likewise, if a neutral chlorine atom (a non-metal) gains an electron, then a negative ion is formed with an overall charge of negative one. The ions above (Na+ and Cl-) form an ionic bond because oppositely charged ions attract each other.
- **Covalent Bonds:** In some chemicals, atoms do not gain or lose electrons but share them. This is called covalent bonding. For example, in methane (CH_4) each carbon (C) and hydrogen (H) share one pair of electrons. Each single line represents a single bond or the sharing of one pair of electrons.

methane

- In ethene (C_2H_4) two pairs of electrons are being shared by the carbons (indicated by the two lines between the carbons). Each single line between a carbon and hydrogen represents a single bond or the sharing of one pair of electrons.

ethene

What Are the Parts of a Chemical?

Holding Together (cont.)

Complete the chart and draw a picture of each model.

Chemical Name	Formula	Number of Single Bonds	Number of Double Bonds	Picture of Model

©1994 Teacher Created Materials, Inc.   #648 Easy Chemistry

What Are the Parts of a Chemical?

Hidden Water

Question

Can unseen water be part of a chemical?

Setting the Stage
- Discuss with students crystals, mentioning that they are solids with definite geometric shapes.
- Discuss with students hydrates (crystals containing water molecules) and water of hydration (water contained in crystals).
- Show students how to use a Bunsen burner.

Materials Needed for Each Group
- approved safety glasses
- two test tubes
- test tube holder
- test tube rack
- Bunsen burner
- sodium chloride (NaCl)
- copper sulfate ($CuSO_4$)
- spatula
- crayons
- data-capture sheet (page 21), one per student

Procedure *(Student Instructions)*
1. Put on a pair of approved safety glasses.
2. Carefully place a few crystals of copper sulfate into a clean, dry test tube. (Use caution with this chemical and ask your teacher how to later dispose of it properly by following the instructions on the Material Safety Data Sheet.)
3. Using the test tube holder, carefully warm the test tube over a small flame. Note what happens.
4. Repeat steps 2 and 3 using sodium chloride.
5. Complete your data-capture sheet.

Extensions
- In the above activity, have students examine the strengths of the bonds of the water to the crystal, compared to the bonds holding the crystal together.
- Have students test other salts to see if they are hydrates.

Closure
- In their chemistry journals, have students write what a hydrate is.
- Have them describe how they broke the bond of water to the crystal.

The Big Why

Hydrates are crystals that contain molecules of water. The presence of water cannot be detected by an ordinary observation of the hydrate. Upon the heating of a hydrate, drops of water form at the top of the test tube. The heat energy breaks the bond holding the water molecules to the crystal. The water contained in the crystal is called water of hydration.

What Are the Parts of a Chemical?

Hidden Water *(cont.)*

Complete the chart and draw pictures showing what the crystal looked like before and after heating. Color with crayons.

Crystal Name	Crystal Picture Before Heating	Crystal Picture After Heating	Hydrate Yes/No

©1994 Teacher Created Materials, Inc. #648 Easy Chemistry

What Are the Parts of a Chemical?

Group Leader

Question

Why is a functional group important to a chemical?

Setting the Stage
- Discuss *organic* chemicals. Tell students that organic chemicals contain carbon.
- Discuss *functional groups*. Tell students that they are specific groupings of atoms that give characteristic properties to the chemicals that they are part of.

Materials Needed for Each Group
- Universal indicator paper
- acetic acid (vinegar)
- sodium hydroxide (baking soda and water)
- plastic teaspoon
- two paper cups
- water
- data-capture sheet (page 23), one per student

Procedure *(Student Instructions)*
1. Fill a paper cup one-third of the way with vinegar, source of acetic acid (CH_3COOH). The organic acid functional group is the COOH group of atoms.
2. Taste the vinegar.
3. Rub the vinegar between your fingers and note the feeling.
4. Using Universal indicator paper, determine the pH of the vinegar sample. (A pH of seven is neutral, values less than seven are acidic, and values greater than seven are basic.)
5. Record all the above data on your data-capture sheet.
6. Place 1 teaspoon (5 g) of baking soda into a paper cup and fill the cup half way with water. This is a source of sodium hydroxide (NaOH). The OH is the base functional group. Be sure to mix the water and baking soda.
7. Repeat steps 2-5 using the baking soda and water.
8. Complete your data-capture sheet.

Extensions
- Have students test other functional groups such as esters. Esters are in banana oil.
- Have students invent their own functional group of a chemical. Tell them to list the properties the functional group gives to the chemical. Have students describe and discuss their functional groups in class.

Closure

In their chemistry journals, have students write some properties a functional group can give a chemical.

The Big Why

The presence of a particular group in a substance determines the substance's properties. For example, the presence of a basic group causes the substance to feel slippery and to have a pH value greater than seven.

What Are the Parts of a Chemical?

Group Leader *(cont.)*

Complete the chart.

Chemical Name	Formula	Functional Group	Taste	Feeling	pH	Acid or Base

What Are the Parts of a Chemical?

Disappearing Act

Question

What happens to radioactive chemicals?

Setting the Stage

- Discuss with students *radioactivity* and *half-life*.
- Tell students that radioactive materials emit particles from the nucleus and that after a half-life, one-half of the original material is left. Emphasize that a half-life can vary from seconds to millions of years.

Materials Needed for Each Group

- paper clips of two different colors (80 of each color)
- stopwatch
- graph paper
- data-capture sheet (page 25), one per student

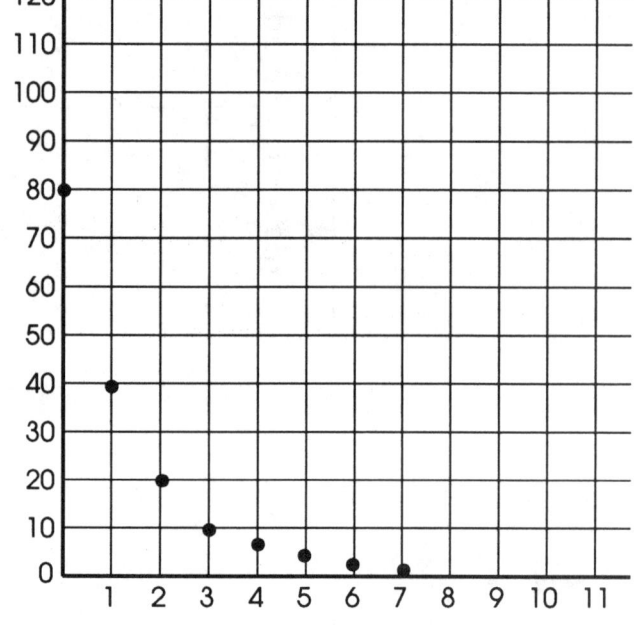

Procedure *(Student Instructions)*

1. Place 80 paper clips of one color on top of the desk. These paper clips represent radioactive material X, which has a half-life of one minute.
2. Using a stopwatch, time a one-minute interval. At the end of the one-minute interval, remove half the paper clips.
3. Again time a one-minute interval and remove half the 40 remaining paper clips.
4. Continue this process until four one-minute intervals have been measured. Note the number of paper clips remaining after each one-minute interval.
5. Place 80 paper clips of the other color on top of the desk. These paper clips represent radioactive material Y with a half-life of two minutes.
6. Repeat steps 2-4 for material Y using two-minute intervals.
7. Complete your data-capture sheet.
8. Using graph paper, plot time in minutes (X axis) versus number of paper clips (Y axis) for radioactive material Y.

Extensions

- Have students extrapolate their graphs (of radioactive material Y) to zero. Have them estimate the amount of time needed for the radioactive material to be all gone.
- Using Reference Tables for Chemistry, compare the half-lives of some radioactive materials.

Closure

In their chemistry journals, have students draw a picture of their graph of radioactive material Y. Have them label it and indicate where each half-life is. Have them note the total number of half-lives included in the graph.

The Big Why

This activity provides a concrete example of what takes place in radioactive material. Students actually see the material disappear.

What Are the Parts of a Chemical?

Disappearing Act *(cont.)*

After performing the activity with the radioactive materials, complete the chart.

Radioactive Material	Half Life (minutes)	Number of Paper Clips Remaining After the Following Number of Minutes						
		1	2	3	4	5	6	7

©1994 Teacher Created Materials, Inc.   #648 Easy Chemistry

What Are the Properties of a Chemical?

Just the Facts

Chemicals contain many properties which are used to help identify them. People are identified and recognized by physical properties or characteristics such as height, hair, and eye color. Brainstorm with your students and together come up with a list of properties to describe a chemical.

Some properties of a chemical are the following:

- **Appearance...**is what a chemical looks like. It includes the chemical's size, shape, taste, color, odor, texture, and state of matter (solid, liquid, or gas).

- **Concentration...**is the amount of a given substance or chemical in a mixture or solution, sometimes expressed by percent of weight or volume.

- **Density...**is the mass per unit volume of a chemical. Keep in mind that volume is the amount of space that the chemical takes up.

- **Surface tension...**is the attractive force in any liquid exerted by the molecules below the surface upon those at the surface.

- **Viscosity...**is a liquid's resistance to flow. Objects move slowly in viscous substances.

What Are the Properties of a Chemical?

Appearance

Question

How does one describe a chemical's appearance?

Setting the Stage

- Discuss *matter* with students. Tell them that matter is anything that has mass and takes up space (volume).

- Discuss with students the *phases* or *states of matter*. Matter may exist as a solid, liquid, or gas. Tell students that solids have a definite shape and volume, liquids have a definite volume and take the shape of their container, and gases do not have a definite shape or volume.

Materials Needed for Each Group

- ten labeled containers of chemicals such as copper wire, zinc, and water.
- magnifying glass
- metric ruler
- data-capture sheet (page 28), one per student

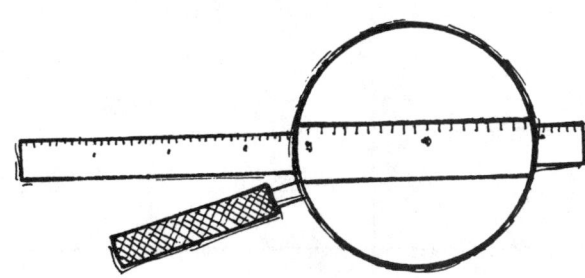

Procedure *(Student Instructions)*

1. Look at each chemical sample using a magnifying glass. Note the appearance.
2. When possible, use a metric ruler to measure the length of chemical samples in centimeters (cm).
3. Complete your data-capture sheet.

Extensions

- Have students analyze the information in the data-capture sheet by noting the similarities and differences in the solid chemical samples.
- Have students be creative and invent their own chemicals.
- Using crayons, have students draw a picture of the chemical. Also have them write a short paragraph describing the chemical. Have students present their chemicals to the class in the form of oral reports. Display the pictures and report on the bulletin board.

Closure

In their chemistry journals, have students list as many ways as they can think of to describe the appearance of a chemical.

The Big Why

In this activity students notice the physical properties of a chemical. They notice the color, shape, size, and phase of the material.

©1994 Teacher Created Materials, Inc. #648 Easy Chemistry

What Are the Properties of a Chemical?

Appearance *(cont.)*

After looking at the chemical samples, complete this chart.

Chemical Name	Phase	Shape	Size (cm)	Color	Other data (odor, etc.)

What Are the Properties of a Chemical?

More than Meets the Eye

Question
Can a chemical contain unseen properties?

Setting the Stage
- Discuss *mixtures* and *solutions* with students. Tell them that *homogeneous* mixtures or solutions (eg., sugar water) are the same throughout. *Heterogeneous* mixtures are not the same throughout (eg., tossed salad).
- Tell students that what they see may not always be all that is present.

Materials Needed for Each Individual
- table salt
- water
- clear plastic cup
- plastic teaspoon
- metric ruler
- marker
- magnifying glass
- data-capture sheet (page 31)

Procedure *(Student Instructions)*
1. Fill a clear plastic cup two-thirds full of water.
2. Add 1 teaspoon (5 g) of table salt to the water and mix.
3. Using the marker, mark the water level on the outside of the cup.
4. Place the cup in a safe place so that it can be checked regularly.
5. Check the cup twice a week for loss of liquid due to evaporation. Mark the level of water each time the cup is checked.
6. Using the metric ruler, measure the difference between the last two marks on the cup to determine the loss of water in cm each time.
7. Continue measuring the loss of liquid (as described above) until all the water has evaporated.
8. Using a magnifying glass, note the contents of the beaker after total evaporation has taken place.
9. Complete your data-capture sheet.

©1994 Teacher Created Materials, Inc.   29   #648 *Easy Chemistry*

What Are the Properties of a Chemical?

More Than Meets the Eye (cont.)

Extensions

- Have students separate the heterogeneous mixture of sulfur and iron filings with a magnet.
- Give students a sample consisting of a mixture of sand and table salt. Have them separate the mixture in the following way. First, place the mixture of sand and salt in water. The salt will dissolve. Then, filter off the sand using filter paper and a funnel. Finally, after evaporation of the water, the salt remains.

Closure

In their chemistry journals, have students record the number of days required to evaporate the water from the plastic cup. Have them list factors that affect the rate of evaporation.

The Big Why

This activity exemplifies the saying that one cannot judge a book by its cover. A chemical may not be properly judged by its outward appearance.

What Are the Properties of a Chemical?

More than Meets the Eye *(cont.)*

Record the cm of water for each day checked. Draw a picture of the contents of the cup after the total evaporation.

Day Checked	Water Loss (cm)	Picture of Cup Contents After Total Evaporation

What Are the Properties of a Chemical?

Solution Dilution

Question
How can the concentration of a chemical solution be changed?

Setting the Stage
- Discuss *concentration* with students. Tell them that it is the amount of a chemical per unit volume. It is usually a certain amount of a solid chemical (called solute) per unit volume of a liquid (called the solvent). The solute is dissolved in the solvent. Together they are called a solution.
- Discuss *dilution* with students. Tell them that a chemical solution can be diluted or made less concentrated by adding water to it.

Materials Needed for Each Group
- labels
- five clear plastic cups
- marker
- data-capture sheet (page 34), one per student
- one-percent milk
- straw
- large plastic bowl

Procedure *(Student Instructions)*
1. Using a marker, draw a horizontal line one-fourth of the way up from the bottom of each plastic cup.
2. Fill one cup to the mark with one-percent milk. Label the cup and set it aside for later use.
3. Fill another cup to the mark with one-percent milk. Pour the measured milk into the plastic bowl.
4. Fill the same cup to the mark with water and pour the measured water into the bowl containing the milk.
5. Continue to measure and pour the water into the bowl eight more times. (Students should have measured one volume of milk and nine volumes of water.)
6. Using the straw, mix the contents of the bowl.
7. Determine the concentration of the diluted milk. It is one-tenth the concentration of the milk used.
8. Fill an empty plastic cup to the mark with contents from the bowl. Label it and set it aside for later use.
9. Fill another empty plastic cup to the mark with the contents from the bowl. (This measured milk will be used in the next dilution.)
10. Empty and rinse out the plastic bowl. Add the contents from step 9.
11. Repeat steps 4 - 10.
12. Repeat steps 4 - 8 for the last dilution.
13. Complete your data-capture sheet.

Extensions
- Have students dilute colored juice or soda and determine the concentration of each dilution.
- Have students make a diluted juice sample more concentrated by using a concentrated juice of the same type and of a known concentration. Tell students the concentration of the original dilute juice sample.

What Are the Properties of a Chemical?

Solution Dilution *(cont.)*

Closure

In their chemistry journals, have students write a few sentences comparing the appearance of the four milk samples. Using their data, have them determine the concentration of a milk sample from a fourth dilution.

The Big Why

This activity allows students to prepare solutions of varying concentrations using water. This skill is very important in medical work. (Medicine of the wrong concentration may be harmful or even fatal.)

What Are the Properties of a Chemical?

Solution Dilution *(cont.)*

Determine the milk concentration after each dilution by multiplying the concentration of the milk used by 0.1. Enter the values below. In the space provided, describe the appearance of each milk sample.

Milk Concentration (%)	Number of Dilutions	Milk Sample Appearance

What Are the Properties of a Chemical?

Density

Question

Why is it important to know a chemical's density?

Setting the Stage

- Discuss *density* with students. Tell them that density is mass per unit volume. For solids, it is mass in grams per cubic centimeter.
- Tell students that a chemical can be identified by its density.

Materials Needed for Each Group

- graduated cylinder (50 mL)
- three different sized rubber stoppers containing a size number and holes in them
- balance
- water
- data-capture sheet (page 36), one per student

Procedure *(Student Instructions)*

1. Using the balance, weigh a rubber stopper to the nearest tenth of a gram. Record on your data-capture sheet.
2. Fill the graduated cylinder with 20 mL of water.
3. Carefully add the rubber stopper to the graduated cylinder containing the water. Note the new volume. Record on your data-capture sheet.
4. Determine the volume of the rubber stopper. The volume of the rubber stopper is found by subtracting the initial volume of 20 mL from the final volume (volume read when rubber stopper is in the graduated cylinder containing the 20 mL of water). Record on your data-capture sheet.
5. Repeat the above steps with the other two rubber stoppers.
6. Complete your data-capture sheet.

Extensions

- Have students use chemistry books and reference tables to look up the densities of various chemicals. Have them write down and compare the densities of several gases. They should be able to make concluding statements about the gases. For example, carbon dioxide gas is more dense than air, so a balloon filled with carbon dioxide gas does not float in air.
- Give students a list of densities and have them identify the chemicals containing those densities. Provide students with a Handbook of Densities.

Closure

In their chemistry journals, have students draw a picture of the graduated cylinder containing 20 mL of water and draw a picture of each of the rubber stoppers in the graduated cylinders of water. Have them label their pictures and write a few sentences comparing the volumes of the three rubber stoppers.

The Big Why

A chemical's density is important in everyday life. Chemical density is taken into consideration in the manufacturing of boats. Boats must be made in such a way that they are less dense than water so that they will float.

What Are the Properties of a Chemical?

Density (cont.)

Record all data below. Determine the density of each rubber stopper by dividing the volume into the mass. Use the following formula: Density = mass in grams divided by the volume in cubic centimeters. Note that 1 mL of water is equal to 1 cubic centimeter.

	Weight in Grams	Volume in cc	Density
Stopper 1			
Stopper 2			
Stopper 3			

What Are the Properties of a Chemical?

Pile It Higher

Question
Why do some chemicals hold together better than others?

Setting the Stage
- Discuss *surface tension* with students. Tell them that surface tension in any liquid is the attractive force exerted by the molecules below the surface upon those at the surface. It is an inward pull which tends to restrain a liquid from flowing.
- Tell students that polar liquids have a higher surface tension than non-polar liquids. A polar liquid such as water has a positive and a negative end.

Materials Needed for Each Individual
- two clear plastic cups
- vegetable oil
- labels
- two pieces of wax paper
- water
- two droppers
- a coin
- data-capture sheet (page 38)

Procedure *(Student Instructions)*
1. Fill one plastic cup halfway with vegetable oil and label it.
2. Fill the other plastic cup halfway with water and label it.
3. Carefully pour a few drops of water on a piece of wax paper. Slowly move the paper around. Note what happens.
4. Carefully pour a few drops of vegetable oil on a piece of wax paper. Slowly move the paper around. Note what happens.
5. Using one of the droppers, count the number of drops of water that can be placed on the head side of a coin.
6. Add dish soap to the plastic cup of water. Using the same dropper, count the number of drops of water with soap that can be placed on the head side of the dry coin.
7. Using the other dropper, count the number of drops of vegetable oil that can be placed on the head side of the coin used in steps 5 and 6. (Clean and dry the coin before doing this step.)
8. Complete your data-capture sheet.

Extensions
- Have students repeat the above activity using heated vegetable oil and water. Have them compare their results with the results of the above activity. Ask students if temperature affects surface tension.
- Have students compare the surface tension of other items.

Closure
In their chemistry journals, have students write a few sentences comparing the surface tension of vegetable oil and water. Also have them tell how the addition of soap to water affected the water's surface tension.

The Big Why
This activity allows students to compare different liquids in terms of their ability to resist flowing. Chemicals that tend to hold together well do not flow easily.

What Are the Properties of a Chemical?

Pile It Higher (cont.)

Record all data and observations below.

Chemical	Appearance on Wax Paper	Number of Drops on Coin	Number of Drops (with soap) on Coin
Water			
Vegetable Oil			

What Are the Properties of a Chemical?

Let's Race

Question

How fast do liquids flow?

Setting the Stage

- Discuss *viscosity* with students. Tell them that viscosity is a liquid's resistance to flow. Objects move very slowly in viscous substances.
- Tell students that the stronger the forces holding the chemical together, the greater the viscosity.

Materials Needed for Each Group

- stopwatch
- a small marble
- shampoo
- soap
- 50 ml graduated cylinder
- water
- molasses
- graph paper
- data-capture sheet (page 40), one per student

Procedure *(Student Instructions)*

1. Pour 50 mL of water into the 50 mL graduated cylinder.
2. Determine how long it takes for a marble to fall to the bottom of the cylinder. Hold the marble over the top of the cylinder.
3. Start the stopwatch at the moment the marble is released. Record the time in seconds on your data-capture sheet.
4. Repeat the above procedure using shampoo.
5. Next clean the marble and graduated cylinder and repeat the procedure using molasses.
6. Clean the marble and graduated cylinder with soap and water.
7. Complete your data-capture sheet.

Extensions

- Using a similar procedure, have students compare the viscosity of three different shampoos.
- Have students bring in an item from home to use for viscosity comparisons. Prepare a class chart or poster listing chemical items and marble fall-times in seconds.
- Have students analyze the chart and list items in order, from lowest viscosity to highest viscosity.

Closure

In their chemistry journals, have students draw a bar graph to represent the amount of time the marbles took to pass through the various liquids. (They may draw the graph on a piece of graph paper and then place it in their journals.)

The Big Why

In this activity students compare the fall-times of a marble in various liquids to study viscosity. Objects move very slowly in viscous substances.

What Are the Properties of a Chemical?

Let's Race (cont.)

Record all data and observations.

Chemical Item	Marble Fall Time (sec.)	Most Viscous Item Tested	Least Viscous Item Tested
Water			
Shampoo			
Molasses			

How Do Chemicals React?

Just the Facts

Chemicals react in many ways. Have the students draw a picture to show a chemical reacting. Tell the students that the picture may illustrate a known or guessed reaction. Have each student share his/her picture and reaction description with the class.

Chemicals react and undergo chemical change to form new substances with new properties. Chemical changes are indicated when any of the following occur: formation of a gas, formation of a precipitate (e.g., solid falling out of solution), a color change, and a gain or loss of heat.

Some of the ways that chemicals react are as follows:

- **Oxidation...**in general, is a reaction in which the element oxygen combines chemically with another substance.

- **Synthesis or combination...**is a chemical reaction in which different substances (chemicals) are added together to form new substances (chemicals).

- **Decomposition...**is a chemical reaction in which a substance breaks down into its component parts.

- **Replacement...**is a chemical reaction in which at least one part of a chemical switches places with a corresponding part of another chemical.

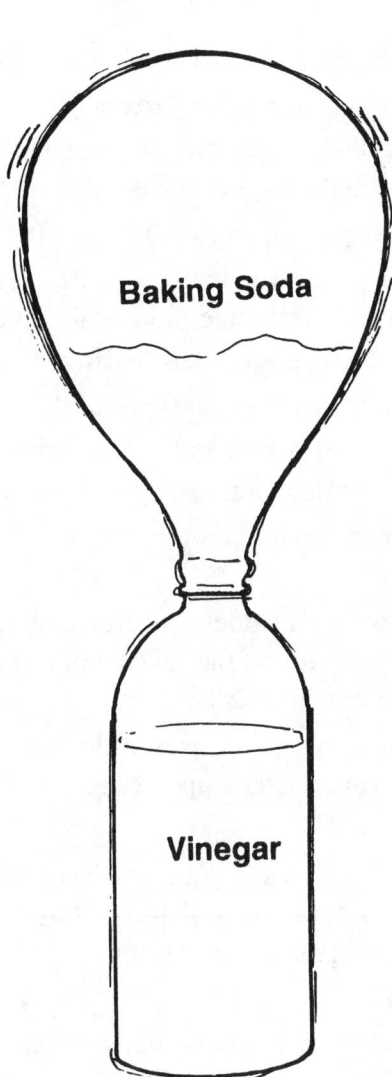

©1994 Teacher Created Materials, Inc. #648 Easy Chemistry

How Do Chemicals React?

Add Oxygen

Question
How do chemicals react with oxygen?

Setting the Stage
- Discuss *chemical reactions* with students. Tell them that a chemical reaction takes place when two or more substances react to form new substances with new properties.
- Discuss *oxidation* with the students. Tell them that in general, oxidation is when oxygen combines chemically with another substance.

Materials Needed for Each Group
- crayon
- apple
- piece of white paper
- water
- potato
- steel wool pad
- plastic knife
- paper cup
- data-capture sheet (page 43), one per student

Procedure *(Student Instructions)*
1. Using the plastic knife, cut the potato in half. Note its appearance and set it aside for later use.
2. Using the plastic knife, cut the apple in half. Note its appearance and set aside for later use.
3. Fill the paper cup two-thirds of the way with water.
4. Note the appearance of a piece of steel wool. Place a piece of it in the cup of water. Set it aside for about 20 minutes and again note its appearance.
5. Place the piece of wet steel wool on a sheet of white paper and let it dry overnight.
6. Again note the appearance of the cut potato and the cut apple.
7. Let the cut potato and the cut apple sit overnight.
8. The next day, note the appearance of the cut potato, the cut apple, and the piece of steel wool.
9. Complete your data-capture sheet.

Extensions
- Discuss with students useful coatings formed by oxidation. For example, *patina* is a corrosion-resisting film on the surface of copper metal statues. It is formed by exposing copper metal to air which contains oxygen.
- Discuss with students non-desirable coatings formed by oxidation. For example, silver ornaments turn greyish-black upon exposure to the air. Describe the use and effect of silver cleaner.

Closure
In their chemistry journals, have students use crayons to draw a picture of their initial observation of each item and of their final observation of each item. In a few sentences have them describe what happened to each item.

The Big Why
This activity reminds students that oxygen is present in the air. It demonstrates what happens when oxygen combines with various substances.

How Do Chemicals React?

Add Oxygen *(cont.)*

In the space provided, describe your observations of the listed items.

Item	Initial Appearance	Appearance of Second Observation	Appearance After Sitting Overnight
Cut Potato			
Cut Apple			
Steel Wool			

How Do Chemicals React?

Blow It Up

Question

How can one detect a chemical change?

Setting the Stage

- Discuss with students ways of detecting chemical change such as the formation of a *gas* or a *precipitate,* a color change or a gain or loss of heat.
- Tell students that a gas can be detected by the formation of bubbles and that a precipitate is indicated by a cloudy solution or by a solid falling out of solution.

Materials Needed for Each Group

- small empty plastic soda bottle with a narrow top
- plastic teaspoon
- vinegar
- data-capture sheet (page 45), one per student
- a regular size balloon
- tape measure
- baking soda

Procedure *(Student Instructions)*

1. Pour vinegar into the soda bottle so that it fills the bottle to a depth of 1" (2.5 cm) from the bottom.
2. Carefully attach the balloon to the top of the bottle. Note what happens.
3. Remove the balloon from the bottle.
4. Add 1 teaspoon (5 g) of baking soda to the inside of the balloon.
5. Carefully re-attach the balloon to the top of the soda bottle. Have the contents of the balloon fall into the bottle. Note what happens.
6. Using a tape measure, make an approximate measurement in centimeters of the circumference of the inflated balloon.
7. Complete your data-capture sheet.

Extensions

- Have students repeat the above activity using different amounts of vinegar and baking soda. Have them record the amounts used. Students can determine the optimum amount of vinegar and baking soda needed to inflate the biggest balloon. Using a tape measure, have students record in centimeters the approximate circumference of each inflated balloon. Display the results on a bulletin board or poster.
- Have students repeat the above activity using balloons of different sizes. Have them determine if one size blows up easier than another.
- Students can simulate lava coming out of a volcano by using vinegar, red food coloring, and baking soda.

Closure

In their chemistry journals, have students describe how they were able to detect a chemical change. Also have them write a sentence or two about what could be done in this activity to increase the size of the inflated balloon.

The Big Why

This activity provides the students with an opportunity to detect a chemical change. Here students detect bubbles which indicate the formation of a gas.

How Do Chemicals React?

Blow It Up (cont.)

In the space provided below, record the data and answer the questions.

Event	Describe what happened to the balloon.	Circumference of Blown Balloon (cm)	Did a chemical change occur? Why?
Balloon & Vinegar			
Balloon, Baking Soda, and Vinegar			

How Do Chemicals React?

Color Change

Question

What can cause a color change in a chemical?

Setting the Stage
- Discuss *indicators* with students. Tell them an indicator is a substance that is one color in an acid environment and a different color in a base environment. Litmus paper is red in acid and blue in base.
- Tell students that an acid has a pH less than seven and that a base has a pH greater than seven. (Lye soap contains a base, and orange juice contains an acid.)

Materials Needed for Each Individual
- Universal indicator paper
- two clear plastic cups
- lemon juice
- baking soda
- plastic teaspoon
- labels
- tea (at room temperature)
- data-capture sheet (page 47)

Procedure *(Student Instructions)*
1. Label one plastic cup tea.
2. Label the other plastic cup tea and lemon juice.
3. Fill each plastic cup halfway with tea.
4. Add about 1 teaspoon (5 g) of lemon juice to the appropriate cup. Stir and note the color. Compare it to the color of the plain tea.
5. Using Universal indicator paper, determine the pH of the contents of each cup. Record all data on your data-capture sheet.
6. Add about 1 teaspoon (5 g) of baking soda to the cup containing lemon juice, until the color closely matches that of the plain cup of tea. Mix and note what happens. (Hint: Bubbles indicate the formation of a gas.)
7. Determine the pH of the contents of the cup containing the baking soda. Record it on your data-capture sheet.
8. Complete your data-capture sheet.

Extensions
- Have students test acids and bases using various indicators such as litmus, phenolphthalein, and others. Make a class chart. In one column list the chemicals tested. Head the other columns with the names of the indicators used. Under these names list the pH values for each chemical.
- Have students use baking soda to test acidic juices such as orange juice, apple cider, and grapefruit juice. Have students first take the pH of each juice. Next have them add and mix about 2 teaspoons (10 g) of baking soda to about one-third of a cup of each juice. Have them again determine the pH of each juice and note any color changes.

Closure

In their chemistry journals, have students list a few ways that they were able to detect a chemical change taking place.

The Big Why

This activity demonstrates color change which indicates a chemical change. In addition to observing different colors, students should note that color represents a particular chemical environment. One color is observed in an acid environment while another color is seen in a base environment.

How Do Chemicals React?

Color Change *(cont.)*

Complete the table below. Also describe the appearance of each item. Include color, reactions taking place, and so on.

Chemical	Initial Appearance	pH	Acid or Base
Plain Tea			
Tea with Lemon Juice			
Tea with Lemon Juice and Baking Soda			

How Do Chemicals React?

Solid Formation

Question
Can a solid be formed by adding two liquids together?

Setting the Stage
- Discuss *precipitates* with students. Tell them a precipitate can be seen as a cloudy solution or as a solid falling out of solution.
- Tell students that the formation of a precipitate indicates that a chemical reaction has taken place.

Materials Needed for Each Group
- four plastic cups
- labels
- two straws
- two coffee filters
- two rubber bands
- red food coloring
- Universal indicator paper
- vinegar
- two-percent milk
- data-capture sheet (page 50), one per student

Procedure *(Student Instructions)*
1. Label the four plastic cups respectively: milk and food coloring, milk with food coloring and vinegar, filter for milk and food coloring, and filter for milk with food coloring and vinegar.
2. Fill the plastic cup labeled milk and food coloring halfway with milk. Add two drops of food coloring and stir with a straw. Note its appearance.
3. Fill halfway with milk the plastic cup labeled milk with food coloring and vinegar. Add two drops of food coloring and stir. Add vinegar to fill this cup three-fourths of the way and stir again. Note its appearance.
4. At this point, use Universal indicator paper to determine the pH of both milk samples and record them on your data-capture sheet.
5. Using rubber bands, place a coffee filter over the top of each unused plastic cup.
6. Slowly pour some of the milk without vinegar over the appropriate filter and let it sit for about five minutes. Observe the results.
7. Slowly pour some of the milk with vinegar over the appropriate filter and let it sit for about five minutes. Observe the results.
8. Feel the top of each filter paper and note the results.
9. Complete your data-capture sheet.

Solid Formation *(cont.)*

Extensions
- Have students determine the mass of a precipitate formed in 10 mL of a milk-vinegar sample using 7 mL of milk and 3 mL of vinegar. First have students weigh a dry, empty coffee filter. Next have students use the weighed coffee filter to filter the 10 mL sample. After collecting all the precipitate on the the filter, let it dry overnight. The next day have the students reweigh the filter containing the precipitate. The mass of the precipitate is the difference between the mass of the filter and the precipitate and the mass of the plain filter paper.
- Have students repeat the above activity using different known amounts of vinegar to milk, to determine what effect the vinegar has on the precipitate formation.

Closure

In their chemistry journals, have students write a few sentences describing this activity. Have them tell whether a chemical reaction took place and explain why.

The Big Why

In this activity students add two liquids together to form a solid. The solid formation indicates that a chemical change has taken place.

How Do Chemicals React?

Solid Formation *(cont.)*

Complete the chart below. Also describe the appearance of the milk samples and the feeling and appearance of the filter paper contents after filtering.

Substance	Appearance	pH	Appearance of Filter Paper Contents	Feeling of Filter Paper Contents
Milk and Food Coloring				
Milk and Food Coloring and Vinegar				

How Do Chemicals React?

Delicious Combination

Question
In what ways do chemicals react?

Setting the Stage
Discuss with students the following ways that chemicals react: *synthesis, decomposition*, and *replacement*. Synthesis occurs when chemicals combine to form one or more new substances. For example, Hydrogen (H) and oxygen (O) combine to form water (H_2O). Decomposition occurs when a chemical breaks down into its component parts. For example, water (H_2O) breaks down into hydrogen (H) and oxygen (O). A replacement reaction occurs when at least one part of one chemical switches places with a corresponding part of another chemical. For example, if one has sodium (Na) and water (H_2O), the sodium (Na) trades places with a hydrogen (H) of water (H_2O) to give sodium hydroxide (NaOH) and hydrogen gas (H).

Materials Needed for Each Group
- recipe for cookies
- ingredients (chemicals) required to make cookies
- bowls, pans, etc. needed to prepare and bake cookies
- data-capture sheet (page 52), one per student

Procedure *(Student Instructions)*
1. Each group should be given a copy of a simple recipe to make cookies. The baking of cookies demonstrates a synthesis reaction. Chemicals are combined to form a new substance. It makes science fun and rewards the students with tasty treats.
2. If possible, do this activity in the school's home economics room where cooking supplies and equipment are available. If that room is not available, do the activity in the school cafeteria.
3. Complete your data-capture sheet.

Extensions
- Obtain a mini-generator and have the students take turns decomposing water by the use of an electric current. This is called *electrolysis* of water. The students should collect twice as much hydrogen as compared to the amount of oxygen gas collected.
- Perform a demonstration for the class. Using a Bunsen burner and a pair of tongs, heat up a piece of magnesium ribbon (Mg). (Wear approved safety glasses.) The magnesium metal in the presence of heat and air gives off light and forms a new substance, a white powder called magnesium oxide (MgO). Have the students compare the appearances of the starting material and the final product.

Closure
In their chemistry journals, have students explain in a few sentences why the baking of cookies is a synthesis reaction. Also have them name and describe the type of energy used to carry out this reaction.

The Big Why
In this activity the students perform a synthesis reaction. Here they combine various ingredients (chemicals) to form a tasty treat (a new substance).

How Do Chemicals React?

Delicious Combination (cont.)

In the space provided, list and describe the ingredients used. Draw a picture of the final product and briefly describe it. (Include taste, appearance, etc.)

Ingredients (Chemicals)	Appearance	State of Matter
1.		
2.		
3.		
4.		
5.		
6.		
7.		

How Do Chemicals React?

Changing Partners

Question

What is formed when one part of a chemical changes places with a corresponding part of another chemical?

Setting the Stage

- Discuss *ions* and *oxidation numbers* with students. Tell them ions are charged atoms of elements. A positive ion is formed when an atom loses one or more electrons, and a negative ion is formed when an atom gains electrons. On the Periodic Table of the Elements, point out the oxidation numbers. (They indicate what ion is formed.) They appear on the upper right hand corner of each elements's square.

- Discuss *formulas* with students. Mention that a formula consists of symbols of elements present in a chemical. The symbol representing the positive part of the substance is usually written first, and the symbol representing the negative part of the substance is usually written last. The overall charge in the formula equals 0. The formula for calcium chloride is $CaCl_2$. Looking at the Periodic Table of the Elements, one sees that calcium (Ca) has an oxidation number of positive two and chlorine (Cl) has an oxidation number of negative one. To write the formula with an overall charge of O, one needs two chlorines to balance calcium's positive two charge. Note that the first part of the formula consists of the positive part of the substance.

Materials Needed for Each Group

- crayons
- six sheets of different colored construction paper
- black marker scissors
- plastic bag
- Periodic Table of the Elements
- data-capture sheet (page 55), one per student

Procedure *(Students Instructions)*

1. Using the Periodic Table of the Elements, look up the following elements' symbols and oxidation numbers: sodium, potassium, magnesium, calcium, oxygen, and chlorine. List them and the ions on your data-capture sheet.

2. On each piece of construction paper, draw a paddle-shaped form to represent one of the six ions. Make two sodium and two chlorine ions. The positive ions should have the circle part of the paddle on the left, and the negative ions should have the circle part of the paddle on the right.

3. Cut out each paddle-shaped form.

4. Using a black marker, label each one. In the circle part of the paddle, write the element's symbol and its oxidation number. Thus the ion is formed.

   Example

5. Using your paddles, combine the following ions. (Hint: Remember that the overall charge in a formula is 0. Therefore, when combining the calcium and chlorine ions, one needs two chlorine paddles and one calcium paddle. Thus the formula for that combination is $CaCl_2$.)

©1994 Teacher Created Materials, Inc.   53   #648 Easy Chemistry

How Do Chemicals React?

Changing Partners (cont.)

Set I   sodium & chlorine
            potassium & chlorine

Set II   calcium & oxygen
            sodium & oxygen

Set III   calcium & chlorine
            magnesium & chlorine

6. Complete your data-capture sheet.
7. When finished, save the paddle forms in a plastic bag for later use.

Extensions

- Have students continue this activity using other elements from the Periodic Table. Have them display the results on a bulletin board.

- Have students do the same type of activity using colored paper clips to represent different elements. Have them hook the paper clips together to combine elements. Then, have them tape and label them on a chart in the classroom.

Closure

In their chemistry journals, have students list three parts of the activity when a changing of partners took place. Have them describe what happens when a chemical changes places with a corresponding part of another chemical.

The Big Why

This activity allows students to simulate synthesis reactions. It provides concrete examples of two elements coming together to form a compound.

1 H 1	12 Mg 2,8,2	20 Ca 2,8,8,2	26 Fe 2,8,14,2	6 C 2,4	7 N 2,5
Hydrogen 1.0079	Magnesium 24.305	Calcium 40.08	Iron 55.847	Carbon 12.011	Nitrogen 14.0067
8 O 2,6	13 Al 2,8,3	14 Si 2,8,4	16 S 2,8,6	17 Cl 2,8,7	28 Ni 2,8,16,2
Oxygen 15.9994	Aluminum 26.9815	Silicon 28.0855	Sulfur 32.06	Chlorine 35.453	Nickel 58.69
29 Cu 2,8,18,1	30 Zn 2,8,18,2	47 Ag 2,8,18,18,1	79 Au 2,8,18,32,18,1	80 Hg 2,8,18,32,18,2	82 Pb 2,8,18,32,18,4
Copper 63.546	Zinc 65.39	Silver 107.868	Gold 196.967	Mercury 200.59	Lead 207.2

How Do Chemicals React?

Changing Partners (cont.)

Complete the charts.

Element Name	Element Symbol	Oxidation Number	Ion	Positive or Negative Ion
ex. calcium	Ca	+2	Ca^{+2}	positive
ex. chlorine	Cl	-1	Cl^{-1}	negative

Combined Ions	Pictures of Paddle Representations of Combined Ions (use crayons)	Formula
sodium and chlorine		
potassium and chlorine		
calcium and oxygen		
sodium and oxygen		
calcium and chlorine	ex.	ex. CaCl$_2$
magnesium and chlorine		

©1994 Teacher Created Materials, Inc.   55   #648 Easy Chemistry

Where Are Chemicals Found?

Just the Facts

Chemicals surround us and allow us to exist. People eat and drink chemicals each day. Ask the students to list several reasons why chemicals are added to food. Discuss these reasons in class. Brainstorm with your students and create a list of places where chemicals are found.

Chemicals are everywhere. They are found in places such as the following:

- The **human body**...contains many chemicals. A large amount of the body contains water. The blood contains the element iron, and the stomach contains hydrochloric acid which helps in the digestive process.

- **Water**...is made from the elements hydrogen and oxygen. It may contain the chemical chlorine, which serves as a disinfectant. Some sources of water contain chemical pollutants.

- **Air**...is a mixture of chemicals called gases. It contains about 78% nitrogen and about 21% oxygen. Air contains chemicals such as water, metallic dust, and other pollutants.

- **Rocks and minerals**...are formed from chemical elements and substances found naturally on the crust of the earth.

- **Food and beverages**...contain chemicals to enhance taste and appearance and to act as preservatives. They contain chemicals such as salts, acids, and gases.

Where Are Chemicals Found?

Human Body

Question
   What are some properties and uses of a gas from the human lungs?

Setting the Stage
- Tell students that the human body is full of chemicals. State that humans breathe in oxygen gas (O) from the air and exhale the waste product, carbon dioxide gas (CO_2).
- Discuss *density* with students. It is a mass per unit volume. Mention that dense substances tend to sink and less dense substances tend to float and rise above denser ones.

Materials Needed for Each Group
- two balloons
- pan
- one sheet of paper
- scale
- water
- data-capture sheet (page 58), one per student

Procedure *(Student Instructions)*
1. Weigh an uninflated balloon to the nearest tenth of a gram. Record its mass on your data-capture sheet.
2. Blow into the above balloon, tie it, and reweigh it. Record its mass. (Some of the gas in the balloon is carbon dioxide.)
3. Determine the mass of the gas in the uninflated balloon. Subtract the mass of the uninflated balloon from that of the inflated balloon. The difference is the mass of the gas. Record this value on your data-capture sheet.
4. Throw the balloon in the air. Note what happens.
5. Place the inflated balloon in a pan of water. Note what happens.
6. Make sounds with the balloon. Blow up a second balloon. Slowly let the gas out and at the same time carefully pull and stretch the opening end of the balloon. Sound is produced from the vibrating gas particles.
7. Let gas do the work. Cut a sheet of paper into small pieces and place them on a desk. After reblowing the second balloon, aim the balloon's opening at the pieces of paper and let the gas escape. Note what happens.
8. Complete your data-capture sheet.

Extensions
- In class, discuss parts of the human body and include the chemicals involved. For example, blood contains iron (Fe) and lots of water (H_2O); the stomach contains hydrochloric acid (HCl) that helps with digestion.
- Have students give oral reports describing how chemicals relate to the human body. Assign a different chemical to each student.

Closure
   In their chemistry journals, have students describe some of the properties of their body's exhaled gas used in this activity. Have them include a comparison of its density with the density of air and the density of water.

The Big Why
   In this activity students see that the human body contains and uses chemicals. The activity shows some properties of the gas carbon dioxide.

©1994 Teacher Created Materials, Inc.

Where Are Chemicals Found?

Human Body (cont.)

Record data and answer questions.

Item	Mass (grams)
Inflated balloon	
Uninflated balloon	
Gas	

Questions

1. What happens when the inflated balloon is thrown in the air? Why?

   _____
   _____
   _____

2. What happens when the inflated balloon is placed in a pan of water?

   _____
   _____
   _____

3. Were sounds made with the balloon? Explain.

   _____
   _____
   _____

4. How was the gas in the balloon used to do work?

   _____
   _____
   _____

Where Are Chemicals Found?

Water

Question

What are some chemicals found in water?

Setting the stage
- To start, tell students that water is made from the elements hydrogen and oxygen.
- Discuss acid rain with students. Mention that pollutants in the air combine with rain to form acid rain which is harmful to plant life.
- Discuss with students, the pH of acids and bases. Acids have a pH less than seven, bases have a pH greater than seven, and water is neutral with a pH of seven. Acids and bases can be harmful to both plant and animal life.

Materials Needed for Each Individual
- water sample
- jar with lid
- label
- Universal indicator paper
- 3"x 5" (7.5 cm x 12.5 cm) piece of colored paper
- marker
- data-capture sheet (page 60)

Procedure *(Student Instructions)*
1. Each student should place an empty jar outside his/her home. The jar should be left outside for a week or two or until about one-half inch of rain water or snow has been collected.
2. Cap the jar, label it (sudent name and address), and take it to school.
3. Using Universal indicator paper, determine the pH of the water sample.
4. Use a marker to write (on the colored paper) the water sample location, its pH, and whether it is neutral, acidic, or basic.
5. All colored pieces of paper containing water sample information should be taped to poster boards and displayed in the classroom. Students should record class data on their data-capture sheets.
6. Complete your data-capture sheet.

Extensions
- Have students repeat the above activity using samples such as tap water, pond water, and so on.
- Discuss with students other chemicals found in water. For example lead (Pb) from pipes is sometimes present in hot tap water. (Lead is soluble in hot water.) Hard water contains calcium carbonate ($CaCO_3$) or calcium sulfate ($CaSO_4$).

Closure

In their chemistry journals, have students write whether the average pH value for the class water samples is acidic, basic, or neutral. Have them explain what this value means.

The Big Why

In this activity students collect water samples from various locations. They determine if the samples contain chemicals that are acids or bases. The activity allows one to locate areas of acid rain.

Where Are Chemicals Found?

Water *(cont.)*

Record the class water sample data and answer the questions.

Water Sample Location	pH	Acid, Base or Neutral

1. Determine the average pH value for the class water samples. (Add all above listed pH values and divide by the number of water samples.) _____

2. Is the average pH value for the class water samples acidic, basic, or neutral? _____

Where Are Chemicals Found?

Air

Question

What are some chemicals found in air?

Setting the Stage
- Discuss with students the main components of air, a mixture of gases. It is about 78% nitrogen (N) and about 21% oxygen (O).
- Tell students that air contains chemicals such as water, metallic dust, and other pollutants.

Materials Needed for Each Group
- balloon
- five cotton balls
- magnifying glass
- labels
- clear plastic cup
- tape
- five small plastic bags with ties
- marker
- data-capture sheet (page 62), one per student

Procedure *(Student Instructions)*
1. Hold an uninflated balloon inside a plastic cup and then blow it up. Note what happens.
2. Observe the space in the cup not occupied by the balloon. This space contains air. Note the air's color, odor, and so on.
3. Obtain particles suspended in air. Use tape to hang five cotton balls in and outside your home.
4. After three days have passed, bring the cotton balls to class. Each cotton ball should be put in a small plastic bag with a label indicating where it had been placed.
5. Use a magnifying glass to examine your collected particles. Note particle properties such as size, color, shape, and odor.
6. Complete your data-capture sheet.
7. Display the air sample particles on the bulletin board.

Extensions
- Do a fog demonstration for the class. Fog is a suspension of liquid droplets in air. Heat a kettle of water until it whistles. At this time, invert a large, clear plastic cup over the steam. The cup becomes foggy. Quickly place it upside down on the desk. Place another identical type of plastic cup upside down next to it. Have students compare the two cups. One is clear, and the other one contains fog.
- Have students do reports on air pollution. Include topics such as smog (a combination of smoke and fog), cigarette smoke, car and industrial exhausts and more.

Closure

In their chemistry journals, have students describe the locations where the least and most amounts of particles were collected from the air.

The Big Why

In this activity students study air pollution. They observe the properties of some chemicals present in the air.

Where Are Chemicals Found?

Air *(cont.)*

Complete the chart and answer the question.

Air Particle Sample's Location	Particle Properties				
	Size	Color	Shape	Odor	Other

Question

1. Based on the balloon activity, describe air's color, odors and so on.

_____

_____

_____

Where Are Chemicals Found?

Rocks and Minerals

Question

What types of chemicals are found in rocks and minerals?

Setting the Stage
- Discuss *minerals* with students. Tell them that minerals are chemical elements such as copper (Cu) or inorganic substances which generally do not contain carbon (C) and are found naturally on the crust of the earth.
- Discuss *rocks* with students. Tell them that rocks are made of minerals and are grouped according to how they are formed. For example, rocks formed from small rock pieces or sediments are called sedimentary rocks.

Materials Needed for Each Group
- carbon (lead of a pencil)
- talc (baby powder)
- sodium chloride (table salt)
- silicon dioxide (sand, glass or quartz)
- aluminum (aluminum foil)
- iron (nail)
- copper (copper wire or a penny before 1980)
- magnet
- magnifying glass
- Rock & Mineral Book
- Periodic Table of the Elements
- two sheets of white paper
- data-capture sheet (page 65), one per student

Procedure *(Student Instructions)*
1. Using a magnifying glass and a magnet, observe, touch, and test the following minerals: carbon, talc, iron, sodium chloride, silicon dioxide, aluminum, and copper. Record the results on your data-capture sheet.
2. Perform a streak test on the above minerals. A streak is helpful in identifying some minerals. Rub each mineral against a sheet of white paper. Record the streak colors on your data-capture sheet.
3. Using a Rock and Mineral book, look up and record the formula for the minerals listed above.
4. Using the Periodic Table of the Elements, identify the elements present in each mineral. Record all data on your data-capture sheet.
5. Complete your data-capture sheet.

Where Are Chemicals Found?

Rocks and Minerals *(cont.)*

Extensions

- Have students determine the hardness range of various minerals using the Mohs' scale. Hardness, a physical property of minerals, is how a mineral resists scratching. The Mohs' scale of hardness goes from 1 to 10 with the softest mineral, talc, having a value of 1 and the hardest mineral, diamond, having a value of 10. A penny has a value of about 3. Using a penny, have the students identify minerals with a hardness less than 3 and those with a hardness greater than 3. (A mineral can be scratched by any other mineral that has a higher number on the Mohs scale.)

- Perform a flame test demonstration for students. Using safety goggles, tongs, and a Bunsen burner, heat a piece of copper wire. (Hold the wire with the tongs.) The copper is identified by its green flame. Next, heat some table salt. (Place wet salt on the ends of the tongs.) Here the sodium is identified by its yellow flame.

Closure

In their chemistry journals, have students list a few ways of obtaining information about minerals' properties.

The Big Why

In this activity students observe the properties of minerals. They also determine the chemical makeup of the minerals.

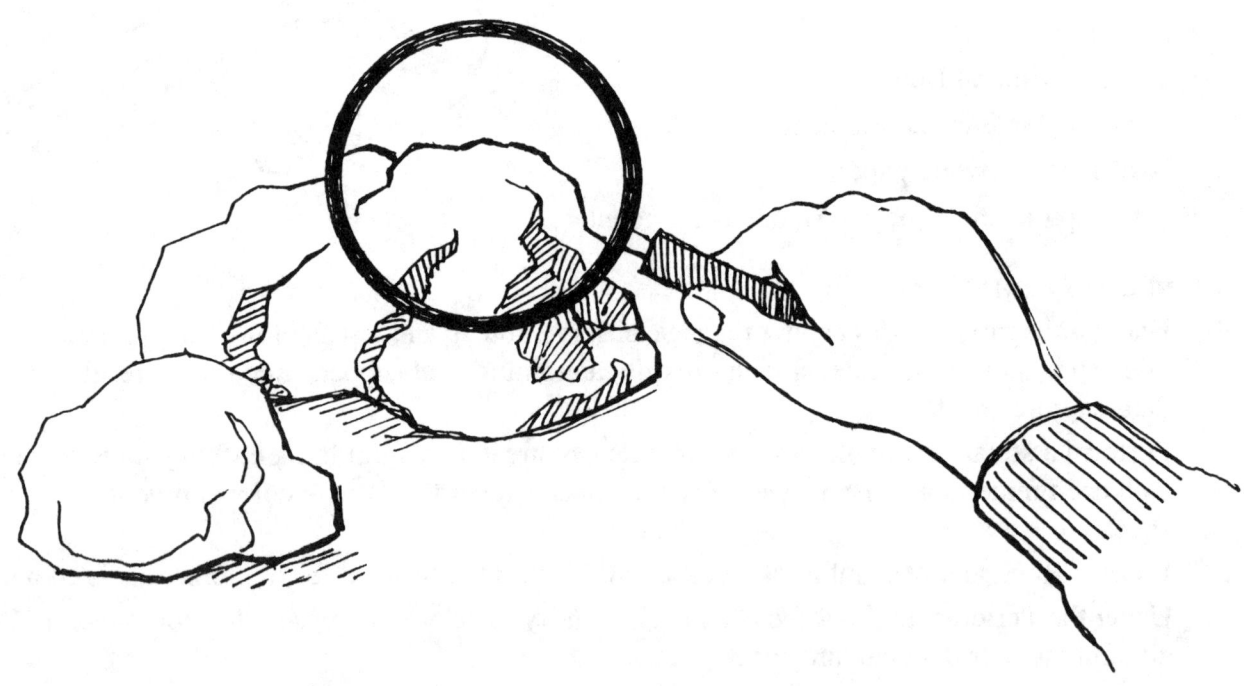

Where Are Chemicals Found?

Rocks and Minerals (cont.)

After examining the mineral samples, complete the table.

Mineral Name	Formula	Elements Present	Sample Color, Shape, etc.	Feeling When Touched	Magnetic Yes/No	Streak Color
1.						
2.						
3.						
4.						
5.						
6.						
7.						

Where Are Chemicals Found?

Food and Beverage

Question

What chemicals are found in food and beverages?

Setting the Stage
- Tell students they eat and drink chemicals every day.
- Mention that some chemicals help preserve food and some enhance its taste.

Materials Needed for Each Group
- five labels or containers of food
- soda
- fruit juice
- powdered fruit drink
- milk
- Universal indicator paper
- Chemical Dictionary
- data-capture sheet (page 68), one per student

Procedure *(Student Instructions)*
1. Bring five labels or containers of food to class.
2. List one chemical contained in each food item.
3. Using a Chemical Dictionary, do the following for each chemical:
   - List the chemical formula.
   - Describe the chemical.
   - Record the use of the chemical in the food item.
4. Repeat steps 2 and 3 for the containers of soda, fruit juice, powdered fruit drink, and milk, provided by the teacher.
5. Using Universal indicator paper, determine and record the pH of the liquids listed in step 4. (A pH of seven is neutral, less than seven is acidic, and greater than seven is basic.)
6. Complete your data-capture sheet.

Extensions
- Have students determine the amount of salt in pretzels. First, have them weigh a piece of paper. Next, have them reweigh the paper with four salted pretzels on it. (Weighings should be done to the nearest tenth of a gram.) The difference in the two weighings is the mass of the salted pretzels. Have the students carefully rub the salt off the pretzels onto the paper and then have them weigh the paper containing all the salt. The difference between the mass of the paper and salt and the plain piece of paper is the mass of the salt. (This activity may be done as a teacher demonstration.)

Where Are Chemicals Found?

Food and Beverage *(cont.)*

- Demonstrate to the class the presence of elemental iron in cereal. Obtain one cup of cereal that contains a large amount of iron. Add about one cup of water to the cereal and beat it to a pulp. Pour the pulp into a beaker, add a magnetic stirrer, and place the beaker on a heating/stirring unit. Note the appearance of iron on the magnetic stirrer.

Closure

In their chemistry journals, have students list different uses of chemicals in food and beverages.

The Big Why

This activity makes students aware of their dependence on chemicals. They eat and drink chemicals each day. Students will begin to analyze food items in terms of their chemical content.

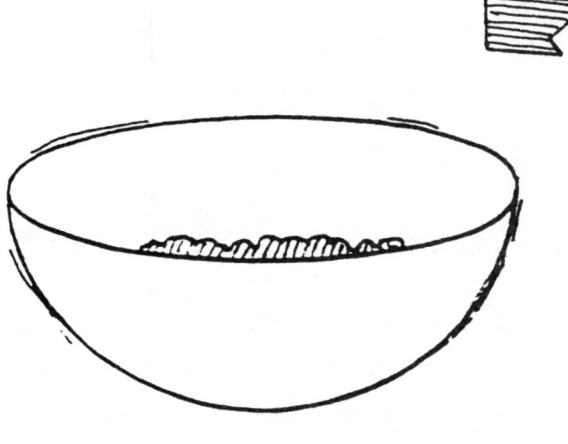

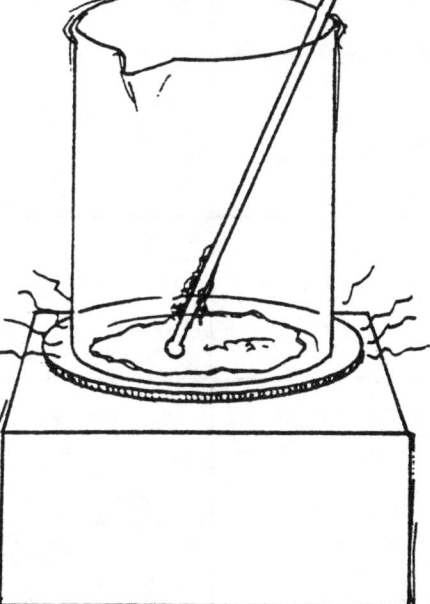

Where Are Chemicals Found?

Food and Beverages *(cont.)*

After analyzing the food and beverages, complete the table.

Food/ Beverage	Listed Chemical	Chemical Formula	Chemical Description	Chemical use in Food/ Beverage	pH

Where Are Chemicals Found?

Everywhere

Question
What items other than food and beverages contain chemicals?

Setting the Stage
Tell students they are surrounded by chemicals. Chemicals are everywhere.

Materials Needed for Each Individual
- crayons
- one item from home to be analyzed
- Universal indicator paper
- data-capture sheet (page 70)

Procedure *(Student Instructions)*
1. Bring an item from home to analyze in class. The item may be shampoo, toothpaste, an antacid, a laundry detergent, or something else.
2. Each student should analyze all the items brought to class and record the data on the data-capture sheet.
3. To analyze each item, do the following:
   - Observe its odor, color, and phase (solid, liquid, or gas).
   - Note what it feels like when touched.
   - Determine its pH, if possible, using Universal indicator paper. (A pH of seven is neutral, less than seven is acidic, and greater then seven is basic.)
   - List at least one chemical in the item.
4. Complete your data-capture sheet.

Extensions
- Have students use a Chemical Dictionary to determine the use of chemicals in the items analyzed.
- Have students compare shampoos in terms of acidity. Good shampoos have a pH close to neutral. Shampoos with an acidic pH can damage one's hair. Have the students rank the shampoos in terms of having a pH closest to neutral. (Students may use Universal indicator paper to determine pH values.)

Closure
In their chemistry journals, have students use crayons to draw a picture of one of the items analyzed. Under their picture, have them write a few sentences describing the properties (color, phase, etc.) of that item.

The Big Why
This activity demonstrates that chemicals are everywhere. Students will begin to analyze a variety of items in terms of chemical content.

Where Are Chemicals Found?

Everywhere (cont.)

After analyzing the items, complete the table.

Item	Odor	Phase: Solid, Liquid, or Gas	Item	Feeling When Touched	pH	Listed Chemical in Item	Other Data

Curriculum Connections

Language Arts

The Language Arts—reading, writing, listening, and speaking can be easily used to teach and reinforce science concepts.

Focus exciting activities involving poems, stories, writing assignments, and dramatic oral presentations around the area of science called chemistry. In general, one can say that chemistry is the study of various substances, their combinations, and the processes by which they act on one another. These substances or chemicals are everywhere: air, water, food, and throughout our world.

Science Concept: *Chemicals are in medicine.*

"News of Medicine," section of *Reader's Digest* issues.

Procedure *(Teacher Instructions)*
1. Have students read the "News of Medicine," section of a *Reader's Digest* issue. (Give each student a different issue of the *Digest*.)
2. Next, have the students make a list of medicines and the chemicals that they contain. They should also include an explanation of how each medicine works.
3. Discuss with students all the medicines in class.

Science Concept: *Chemicals are in food.*

Ponte, Lowell. "Foods That Help You Live Longer," *Reader's Digest*, June 1993: 150 - 154.

Naj, Amal. "Peppers: Hot, Hotter, Hottest," *Reader's Digest*, July 1993 : 115 - 117.

Both stories describe various foods and the chemicals that they contain. Some disease-preventing ingredients of specific foods are discussed.

Procedure *(Teacher Instructions)*
1. Have students read the above stories aloud in class.
2. Afterwards, have the students prepare a chart containing the following information: food items, chemicals in each food item, and each chemical's use.
3. Discuss with students the information in class.
4. Ask students if they eat any of the food items discussed.

Curriculum Connections

Social Studies

Chemicals have played a significant role in history. Civilizations have thrived or declined because of them, cultures have been built around their use, and people have devoted their lives to working with them to make conditions in the world better in some way.

As you guide your students through lessons in history, geography, cultural awareness, or other areas of social studies, keep in mind the role chemicals have played. You will find it easy to incorporate the teaching and reinforcement of science concepts in your lessons.

Science Concept: *Chemicals are important around the world.*

- Have students brainstorm to obtain a list of chemicals and the products made from those chemicals. Have the students look up the listed items in books such as encyclopedias to find their locations throughout the world. For example, iron, which is used to make steel, is found in places such as North America, South America, and Africa. Make a class chart of chemicals, chemical products, and locations. Display it.

- Have students write and present oral reports on an assigned explorer. Explorers have traveled the world in search of riches such as the chemicals gold (Au) and silver (Ag). Reports should include the explorer's name, country sailed for, date, place, and chemicals or chemical products found.

- Have class investigate wars that have been fought over chemicals. For example, wars have been fought over oil, a source of fuel, located in the Middle East and North Africa. Oil is a chemical that consists mainly of the elements hydrogen (H) and carbon (C). Discuss the information in class.

- Do a class project on currencies of the world. Each student, assigned a different currency, should prepare a chart that includes a currency description and chemical makeup as well as the country using it. For example, U.S. pennies made before 1982 contain more copper (Cu) than pennies made after 1982. Pennies made after 1982 contain much zinc (Zn). Display the charts in the classroom.

 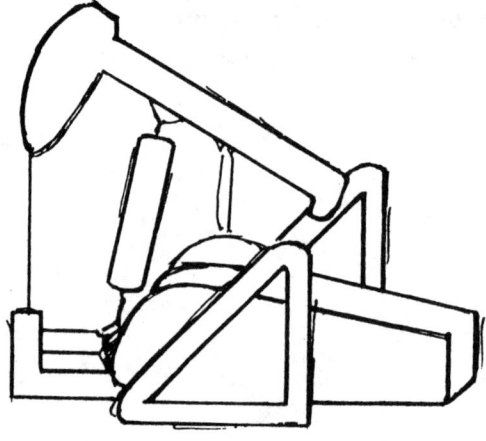

Curriculum Connections

Physical Education

Chemistry is very much a part of physical education. In order to walk, run, and exercise, one breathes in oxygen (O) and exhales the waste product carbon dioxide (CO_2).

Science Concept: *Human life depends on chemicals.*

Aerobic exercises are activities that make the heart beat faster, increase oxygen to the muscles, and bring more air to the lungs. These exercises are excellent for promoting heart and blood vessel fitness.

- If a pool is available, have students swim for an aerobic exercise.

- Another good aerobic exercise is walking. Take students on a walk. Have them make a list of chemicals noticed on the walk. For example, chlorophyll is the chemical that gives grass its green color. They should also note air (mixture of gases), water, car exhaust, etc. Later, discuss the chemical lists in class.

- A third good aerobic exercise is doing aerobics. This can be done in class with a tape and a televison. Both can be rented.

NOTE: When one runs or exercises, a chemical reaction takes place in the muscles. If one does intense exercise and not enough oxygen is present, then the chemical lactic acid gathers in the muscle cells and causes muscle fatigue.

©1994 Teacher Created Materials, Inc.   73   #648 Easy Chemistry

Curriculum Connections

Math

Math and science are very closely related. Math is the language of science, whereas science is very often expressed in mathematical terms. Also science experiments involve the mathematical skills of calculating, weighing, and measuring.

To celebrate a special holiday or reward a good class, do the following activity involving chemistry and math.

Science Concept: *Mathematical skills are used in the chemical reaction called synthesis. (A synthesis reaction occurs when chemicals, each in a set mathematical amount, are combined to form a new product).*

Preparation of Tasty Treats

In this activity, students combine chemicals (ingredients) to form a new product (tasty treat). They demonstrate mathematical skills by modifying a given recipe and by measuring and combining ingredients. Have the students perform this activity in the school cafeteria or in the home economics room where ovens and cooking utensils are available.

Procedure *(Teacher Instructions)*

1. Give each student a copy of a favorite recipe (special cookie, cake etc.).
2. Tell students to calculate the amount of each ingredient required to cut the recipe in half. (If the original recipe called for 1 cup (250 g) of sugar, for example, the modified one would require ½ cup (125 g) of sugar).
3. Have students rewrite the recipe using the newly calculated ingredient amounts.
4. Divide the class into groups consisting of two or three students. Have each group use the modified recipe to prepare the treats. (Have members of each group work cooperatively and take turns measuring and combining ingredients.)
5. Enjoy!

Curriculum Connections

Art

The tools of art relate to chemistry. In art class, one sketches with pencils, colors with crayons, and paints with paints. The lead in pencils is really carbon (C). Crayons contain chemicals such as pigments or dyes in a wax or oil medium. Paint contains chemicals, usually a mixture of pigments and water or oil.

Science Concept: *A color may be separated into its component parts (chemicals).*

Color Separation

Materials Needed for Each Group
- two toothpicks
- a paper cup
- a white paper towel
- paper
- scissors
- red and blue food coloring
- water

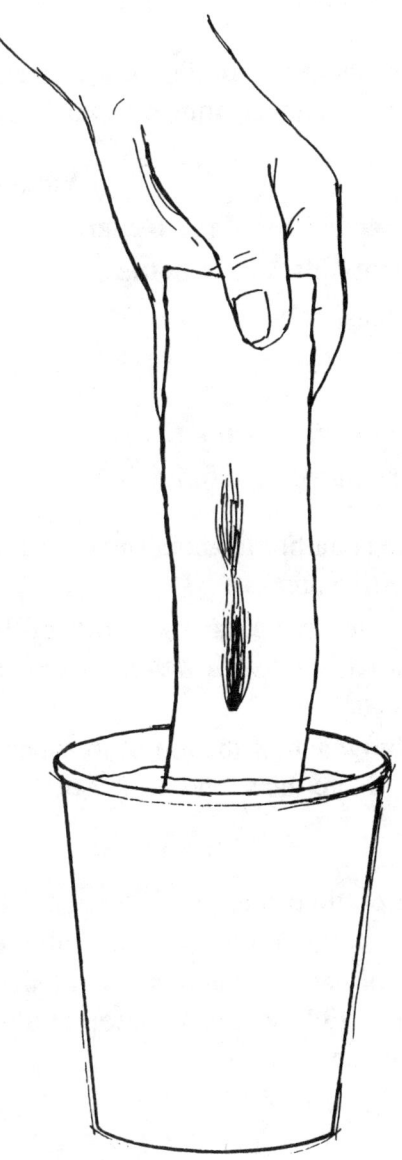

Procedure *(Teacher Instructions)*
Have each student do the following:

1. Cut the paper towel into strips about 1" x 3" (2.54 cm x 7.5 cm) long.
2. Fill a paper cup two-thirds of the way with water.
3. Place one drop of red food coloring on a piece of paper.
4. Next, place one drop of blue food coloring on top of the red drop.
5. Mix the two drops of food coloring together with a toothpick.
6. Place the tip of a clean toothpick into the mixed food coloring and use it to make a dark-colored dot in the center of the paper towel strip.
7. Hold the paper towel strip in the cup of water in such a way that the dot is not in the water.
8. Note the color separation.

©1994 Teacher Created Materials, Inc. 75 #648 Easy Chemistry

Curriculum Connections

Music

Music and chemistry are closely related fields. Musical instruments are made of chemicals. Brass instruments such as trumpets and trombones are made of a copper (Cu) and zinc (Zn) alloy. (An alloy is a substance with metallic properties and consists of two or more elements.) A number of songs have been written about chemicals. "The Blue Danube," by Johann Strauss and "Waves of the Danube," by J. Ivanovici are about the chemical water (H_2O). The waltz, "Gold And Silver," by Franz Lehar contains chemicals in its title. Musical sounds can be produced by the vibration of chemicals. Stringed instruments produce sounds by vibrating strings of different metals, and brass instruments produce sounds as a result of vibrating columns of air (mixture of chemicals known as gases).

Science Concept: *Vibrating chemicals produce musical sounds.*

Have a guest speaker describe how instruments are made and how they produce sounds. (The speaker may be a music teacher, someone who repairs instruments, or an employee of a music store.)

Vibrating Chemicals Activity

Materials Needed for Each Group
- one empty glass soda bottle
- metal teaspoon
- water

Procedure *(Teacher Instructions)*
Have the students do the following:

1. Fill the soda bottle about one-fourth of the way with water.
2. Using a teaspoon, tap the center of the bottle in an area filled with water. Note the sound produced.
3. Next blow across the top of the uncapped bottle. Note the sound produced.

The Big Why
Tapping the bottle produces a high pitched sound due to a short column of vibrating water (H_2O). Blowing across the top of the bottle produces a lower pitched sound due to a long column of vibrating air (mixture of gases). Pitch is the highness and lowness of sound. To extend this activity, have the students tap and blow across the top of a bottle filled three-fourths of the way with water. Discuss the sounds produced.

Teacher Note: See page 10 for instructions.   *Station-to-Station Activity*

Observe

Before beginning your investigation, write your group members' names on the lines below.

_____Team Leader   _____Chemist

_____Stenographer   _____Transcriber

Work together using your senses and a magnifying glass to gather information about these chemicals. Remember that a liquid takes the shape of its container and that a chemical can have the following phase forms—solid, liquid, or gas. Record all data in the space provided. When you have finished, on the back of this activity sheet draw a picture and write a brief description of the group's favorite chemical.

Water
(H_2O)

shape: _____
color: _____
smell: _____
phase: _____
other: _____

Table Salt
sodium chloride (NaCl)

shape: _____
color: _____
smell: _____
phase: _____
other: _____

Sand
silicon dioxide (SiO_2)

shape: _____
color: _____
smell: _____
phase: _____
other: _____

Baking Soda
sodium bicarbonate ($NaHCO_3$)

shape: _____
color: _____
smell: _____
phase: _____
other: _____

Put your finished activity paper in the collection pocket on the side of the table at this station.

©1994 Teacher Created Materials, Inc.   #648 Easy Chemistry

Station-to-Station Activity   **Teacher Note: See page 10 for instructions.**

Communicate

Before beginning your investigation, write your group members' names on the lines below.

_____Team Leader   _____Chemist

_____Stenographer   _____Transcriber

On this table you will find eight chemicals to use to make a circle graph. This graph will communicate at a glance how many solids, liquids, and gases you have grouped together.

1. Sort the chemicals into three groups—solids, liquids, or gases.

2. Decide on a different color to represent each group. Fill in the color key near the circle graph.

3. Count the number of solids, liquids, and gases.

4. In your chosen color for solids, color the same number of circle sections as there are solids.

5. Color the liquid and gas sections in the same way.

6. Determine and list the fraction of each type of chemical present. (For example, the total number of chemicals present is eight. If two of the chemicals are solids, then the fraction of solids is two over eight, or in reduced form it is one-fourth.)

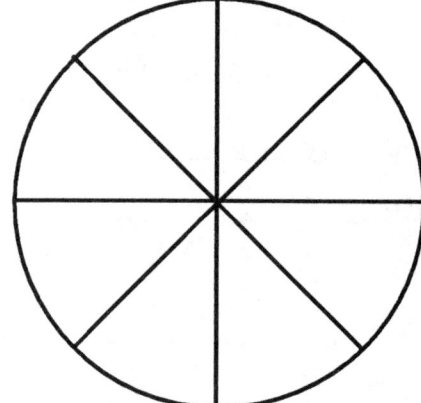

COLOR KEY

Solids = [   ]

Liquids = [   ]

Gases = [   ]

Put your finished activity paper in the collection pocket on the side of the table at this station.

#648 Easy Chemistry   ©1994 Teacher Created Materials, Inc.

Teacher Note: See page 10 for instructions. *Station-to-Station Activity*

Compare

Before beginning your investigation, write your group members' names on the lines below.

_____Team Leader   _____Chemist

_____Stenographer   _____Transcriber

Examine the chemicals at this station.

Chemical 1—water

Chemical 2—baking soda in water

Chemical 3—vinegar

Use Universal indicator paper to determine the pH of each chemical. (A pH value of less than seven is acidic, a value greater than seven is basic, and a value of seven is neutral.) Record the values on the bar graph below. (For each chemical, color the proper number of rectangles).

pH VALUES OF CHEMICALS

	Chemical 1	Chemical 2	Chemical 3
13			
12			
11			
10			
9			
8			
7			
6			
5			
4			
3			
2			
1			
0			

Put your finished activity paper in the collection pocket on the side of the table at this station.

©1994 Teacher Created Materials, Inc.   #648 Easy Chemistry

Station-to-Station Activity Teacher Note: See page 10 for instructions.

Order

Before beginning your investigation, write your group members' names on the lines below.

_____Team Leader   _____Chemist

_____Stenographer   _____Transcriber

There are ten chemicals on this table. The chemicals are present in different amounts. Arrange them in order of smallest to largest amount. In the space provided, write the names of the chemicals in the order you have made. Work together to think of a different way the chemicals can be ordered (eg., from largest to smallest amount). Write your idea and name on the back of this activity sheet.

Smallest to Largest

1	2	3	4	5

6	7	8	9	10

Put your finished activity paper in the collection pocket on the side of the table at this station.

Teacher Note: See page 10 for instructions.   *Station-to-Station Activity*

Categorize

Before beginning your investigation, write your group members' names on the lines below.

_____Team Leader   _____Chemist

_____Stenographer   _____Transcriber

On the table you will find a variety of chemical items such as nails, coins, ingredients in food, and so on. Test each item with a magnet. Then categorize them on the charts below by listing the chemical items under the appropriate headings. Hint: Some items may be listed under more than one heading.

Categorizing Chemical Items

Chemical Item	Food Ingredient	Silver Color	Copper Color	Magnetic	
chocolate chip	✓				

Put your finished activity paper in the collection pocket on the side of the table at this station.

©1994 Teacher Created Materials, Inc.

Station-to-Station Activity   Teacher Note: See page 10 for instructions.

Relate

Before beginning your investigation, write your group members' names on the lines below.

_____Team Leader   _____Chemist

_____Stenographer   _____Transcriber

At this table you will find chemicals or pictures of chemicals and their products. Use what you know about chemicals to relate each chemical to its product. For example, the chemical iron is used to make nails. Make drawings of the matching pairs of chemicals and chemical products. Label each picture with identifying words.

On the back of this activity page, draw and label one more chemical and chemical product pair.

chemical	chemical product	chemical	chemical product
chemical	chemical product	chemical	chemical product

Put your finished activity paper in the collection pocket on the side of the table at this station.

#648 Easy Chemistry   82   ©1994 Teacher Created Materials, Inc.

Teacher Note: See page 10 for instructions.   Station-to-Station Activity

Infer

Before beginning your investigation, write your group members' names on the lines below.

_____Team Leader   _____Chemist

_____Stenographer   _____Transcriber

You will find containers of six different chemicals at this station (flour, water, salt in water, sulfur, iron filings, and plain table salt). It is your job to use what you know about chemicals to write each chemical's name in the appropriate place on this page.

Precipate	Mixture
Solution	Compound

Put your finished activity paper in the collection pocket on the side of the table at this station.

©1994 Teacher Created Materials, Inc.   #648 Easy Chemistry

Station-to-Station Activity **Teacher Note: See page 10 for instructions.**

Apply

Before beginning your investigation, write your group members' names on the lines below.

_____Team Leader   _____Chemist

_____Stenographer   _____Transcriber

At this station is a sample consisting of a mixture of sand, table salt, and iron filings. Apply your knowledge of chemistry and design a plan to separate the components of this mixture. Write your plan in the space provided. Include needed supplies.

Use the materials provided at this station to carry out your plan.

Separation Plan

Put your finished activity paper in the collection pocket on the side of the table at this station.

Management Tools

Science Safety

Discuss the necessity for science safety rules. Reinforce the rules on this page or adapt them to meet the needs of your classroom. You may wish to reproduce the rules for each student or post them in the classroom.

1. The use of dangerous chemicals should be avoided if at all possible, but if necessary should be handled only by the teacher. When ordering chemicals, ask for Material Safety Data Sheets (MSDS) and refer to them when using the chemicals. These sheets contain the information a user needs on potential hazards of each chemical and the methods to be considered for the use and handling of the material. Fire extinguishers and first-aid kits should always be available. Never leave the classroom when doing science experiences.

2. Begin science activities only after all directions have been given.

3. Never put anything in your mouth unless it is required by the science experience.

4. Always wear safety goggles when participating in any lab experience.

5. Dispose of waste and recyclables in proper containers.

6. Follow classroom rules of behavior while participating In science experiences.

7. Review your basic class safety rules every time you conduct a science experience.

You can still have fun and be safe at the same time!

Management Tools

Chemistry Journal

Chemistry Journals are an effective way to integrate science and language arts. Students are to record their observations, thoughts, and questions about past science experiences in a journal to be kept in the science area. The observations may be recorded in sentences or sketches which keep track of changes both in the science item or in the thoughts and discussions of the students.

Chemistry Journal entries can be completed as a team effort or an individual activity. Be sure to model the making and recording of observations several times when introducing the journals to the science area.

Use the student recordings in the Chemistry Journals as a focus for class science discussions. You should lead these discussions and guide students with probing questions, but it is usually not necessary for you to give any explanation. Students come to accurate conclusions as a result of classmates' comments and your questioning. Chemistry Journals can also become part of the students' portfolios and overall assessment program. Journals are valuable assessment tools for parent and student conferences as well.

How To Make a Chemistry Journal

1. Cut two pieces of 8.5" x 11" (22 cm x 28 cm) construction paper to create a cover. Reproduce page 87 and glue it to the front cover of the journal. Allow students to draw chemistry pictures in the box on the cover.
2. Insert several Chemistry Journal pages. (See page 88.)
3. Staple together and cover stapled edge with book tape.

Management Tools

My Chemistry Journal

Name _____

Management Tools

Chemistry Journal

Illustration

This is what happened: _____

_____

_____

_____

_____

This is what I learned: _____

_____

_____

_____

_____

Management Tools

Investigation Planner

Observation

_____

_____

Question

_____

_____

Hypothesis

_____

_____

Procedure

   Materials Needed:

_____

_____

_____

_____

   Step-by-Step Directions: (Number each step.)

_____

_____

_____

_____

_____

_____

©1994 Teacher Created Materials, Inc.   #648 Easy Chemistry

Management Tools

Chemistry Observation Area

In addition to station-to-station activities, students should be given other opportunities for real-life science experiences. For example, models of atoms and simple chemical reactions can provide a vehicle for discovery learning if students are given enough time and space to observe them.

Set up a chemistry observation area in your classroom. As children visit this area during open work time, expect to hear stimulating conversations and questions among them. Encourage curiosity but respect their independence!

Books with facts pertinent to the subject, item, or process being observed should be provided for students who are ready to research more sophisticated information.

Sometimes it is very stimulating to set up a science experiment or add something interesting to the chemistry observation area without a comment from you at all! If the experiment or materials in the observation area should not be disturbed, reinforce with students the need to observe without touching or picking up.

Management Tools

Assessment Form

Cooperative Group Evaluation

Assignment: _____

Date: _____

Scientists	Jobs
_____	_____
_____	_____
_____	_____

As a group, decide which face you should fill in and complete the remaining sentences.

1. We finished our assignment on time, and we did a good job.

2. We encouraged each other, and we cooperated with each other.

3. We did best at _____

   _____.

4. Next time we could improve at _____

   _____.

©1994 Teacher Created Materials, Inc.

Management Tools

Glossary

Acid—a chemical with a sour taste and a pH value less than seven.
Atom—the smallest particle that describes an element. The center of the atom contains protons and neutrons. Electrons move around the center of the atom.
Atomic Number—the number of protons in the atom of an element.

Base—a chemical with a slippery feeling, bitter taste, and a pH value greater than seven.
Bohr model—a model of the atom where electrons revolve around the nucleus in concentric orbits.
Bond—holds something together, an attractive force between atoms.

Chemical Change—the formation of a new substance with new properties.
Chemical Reaction—a chemical change that may take place in several ways such as synthesis, replacement, or decomposition.
Compound—a substance made of atoms or ions of two or more elements that have been combined chemically.
Concentration—a high level of a specified substance in another substance.
Conclusion—the outcome of an investigation.
Control—a standard measure of comparison in an experiment. The control always stays constant.
Covalent Bonding—a compound in which atoms share electrons equally.

Density—the property of a substance equal to its mass per unit volume.

Electrolysis—the decomposition of water by means of an electric current.
Electron—a negatively charged particle (charge of negative one) that moves around the nucleus of the atom.
Element—a substance that can not be broken down into a simpler substance.
Experiment—a means of proving or disproving an hypothesis.

Formula—it contains the symbols of the elements in a chemical.
Functional group—a specific group of atoms that gives characteristic properties to the chemical that they are part of.

Half-life—the amount of time required for one-half of the original radioactive material to exist.
Homogenous Mixture—a mixture that is the same throughout.
Heterogeneous mixture—a mixture that is not the same throughout.
Hydrate—a crystal containing water molecules.
Hypothesis—an educated guess to a question which you are trying to answer.

Indicator—a dye that changes color in the presence of certain substances.
Investigation—observation of something followed by a systematic inquiry in order to explain what was originally observed.
Ion—a charged atom.
Ionic bonding—bonding formed by oppositely charged ions attracting each other.

Litmus paper—an indicator that is red in acid and blue in base.

Matter—anything that has mass and takes up space.
Molecule—a group of bonded atoms that exist as a separate entity.
Neutron—neutral particle in the nucleus or center of an atom.

©1994 Teacher Created Materials, Inc.   93   #648 Easy Chemistry

Glossary (cont.)

Nonpolar—a substance without a positive and negative end (no poles).
Nucleus—the center of the atom that contains the protons and neutrons.

Observation—careful notice or examination of something.
Organic—pertaining to life.
Oxidation—oxygen combining chemically with another substance.
Oxidation number—pertains to the number of electrons an atom gains or loses.
Oxygen—an element that is a gas and necessary for human life.

pH—value used to identify acids and bases. A pH value less than seven is acidic, a value of seven is neutral, and a value greater than seven is basic.
Phases—the states or forms of matter: solid, liquid, and gas.
Polar—a substance with a positive and negative end.
Precipitate—a solid falling out of a solution or seen as a cloudy solution.
Procedure—the series of steps that is carried out when doing an experiment.
Proton—a particle with a charge of a positive 1 and located in the center of an atom.

Question—a formal way of inquiring about a particular topic.

Radioactive material—a material that emits particles from its nucleus.
Replacement—a chemical reaction in which at least one part of a chemical switches places with a corresponding part of another chemical.
Results—the data collected after performing an experiment.

Scientific Method—a creative and systematic process of proving or disproving a given question, following an observation. Observation, question, hypothesis, procedure, results, conclusion, and future investigations.
Scientific-Process Skills—the skills necessary to have in order to be able to think critically. Process skills include: observing, communicating, comparing, ordering, categorizing, relating, inferring, and applying.
Solvent—a substance in which another substance (the solute) is dissolved.
Surface Tension—an invisible film on the surface of liquids, causing them to stick together
Synthesis—a chemical reaction in which different substances are added together to form new substances.

Variable—the changing factor of an experiment.
Viscosity—a liquid's resistance to flow.
Volume—the amount of space something takes up. Mathematically it is determined by multiplying together any item's width, length, and height.

Bibliography

Agler, Leigh. *Liquid Explorations.* Lawrence Science, 1987.

Arms, Karen. *Environmental Science.* Saunders College Publishing, 1990.

Bacher, Angela, Dean Hurd, Charles McLaughlin, and Myrna Silver. *Physical Science.* Prentice Hall, 1988.

Barry, Dr. Dana. *Innovative Activities for Elementary and Middle School Science.* Burgess Publishing Company, 1992.

Barry, Dr. Dana and James Barry. *Modern Chemistry Experiments.* Dana and Jim Barry Inc., 1992.

Barry, Dr. Dana. *Fat Burgers, Science Scope.* NSTA, September 1990: 34 -36.

Barry, Dr. Dana. *Delicious Matter, Science Scope.* NSTA, January 1990: 25.

Barry, Dr. Dana. *Interest Bearing Coins, Science Scope.* NSTA September 1989: 18 - 21.

Brown, Theodore and Eugene LeMay. *Chemistry:The Central Science.* Prentice Hall, 1988.

Campbell, Neil. *Biology.* The Benjamin/Cummings Publishing Company, 1990 ed.

Chesterman, Charles W. *The Audubon Society Field Guide to North American Rocks and Minerals.* Alfred A. Knopf, Inc., 1990 ed.

Chisholm, Jane and Mary Johnson. *Introduction to Chemistry.* Usborne Publishing Ltd., 1983.

Cobb, Vicki. *Chemically Active! Experiments You Can Do At Home.* Harp C Books, 1985.

Cobb, Vicki. *Gobs of Goo.* Harp C Child Bks, 1983.

Ebbing, Darrell. *General Chemistry.* Houghton Mifflin Co., 1990 ed.

Freeman, Ira Mae. *The Story of Chemistry.* Random, 1962.

Gabriel, Lucretia, James McGuirk, Clifford Phillips, William Ramsey and Frank Watenpaugh. *General Science.* Holt, Rinehart and Winston, Publishers, 1988.

Gold, Carol (Editor). *Foodworks.* Addison-Wesley Publishing Company, Inc., 1987.

Hazen, Robert and James Trefil. *Science Matters.* Doubleday, 1991.

Jennings, Terry. *Everyday Chemicals.* Childrens, 1989.

Johnson, May. *Chemistry Experiments.* EDC, 1983.

Kramer, Stephan. *How to Think Like a Scientist: Answering Questions by Scientific Method.* Harp C Child Bks, 1987.

Landau, Elaine. *Chemical & Biological Warfare.* Dutton Child Books, 1991.

Lewis, Richard and N. Irving Sax. *Hawley's Condensed Chemical Dictionary* (eleventh ed.). Van Nostrand Reinhold Company, 1987.

Martin, Paul D. *Science: It's Changing Your World.* Natl. Geog., 1985.

Naj, Amal. *Peppers: Hot, Hotter, Hottest, Reader's Digest,* July 1993: 115 - 117.

Palder, Edward. *Chemistry Magic.* Woodbine Hse, 1987.

Parker, Steve. *Chemistry.* Watts, 1990.

Ponte, Lowell. *Foods That Help You Live Longer, Reader's Digest,* June 1993: 150 - 154.

Science and Children, NSTA. (all issues)

Science Scope, NSTA. (all issues)

Radford, Don. *Looking at Metals.* David & Charles, 1985.

Reymond, Jean Pierre. *Metals: Born of Earth & Fire.* Young Discovery Lib., 1988.

Bibliography *(cont.)*

Silverstein, Alvin. *Vitamins and Minerals.* Milbrook Press, 1992.

Simon & Schuster Staff. *Why Things Are: A Guide to Understanding the World Around Us*. S & S Trade, 1988.

Watson, Philip. *Liquid Magic.* Morrow, 1983.

Whyman, Kathryn. *Chemical Changes.* Watts, 1986.

Whyman, Kathryn. *Metals & Alloys.* Watts, 1988.

Zumdahl, Steven S. *Chemistry* (second ed.). D.C. Heath and Company, 1989.

Technology

Laser Disc. *Matter And Energy For Beginners*, Vol. 1. Cornet Films, 1990 (The laser disc has two titles—Solid, Liquid, Gas and How Materials Change.)

Matter And Energy For Beginners, Vol. 3. Cornet Films, 1990. (The laser disc has two titles—Sound and Light.)

Video Cassette. *Atmosphere: On the Air.* National Geographic Society, 1993.

Chemistry Matters. Disney, 1986.

Measurement: Foundation of Chemistry. (World of Chemistry Series) Annenberg/CPB, 1989.

The Atom. Annenberg/CPB, 1989.

The Water Cycle. (Science Essentials: Oceans and Space) Encyclopedia Britannica Educational Corporation, 1991.

The World of Chemistry Introduction. (World of Chemistry Series) Annenberg/CPB, 1989.

What Are Solids, Liquids And Gases? (Science Essentials: Learning About Matter) Encyclopedia Britannica Educational Corporation, 1991.

What Is a Gas? Encyclopedia Britannica Educational Corporation, 1989.

What is a Liquid? Encyclopedia Britannica Educational Corporation, 1989.

What is a Solid? Encyclopedia Britannica Educational Corporation, 1989.